Waste Disposal Systems in Slaughterhouses Suitable for Developing Countries

Waste Disposal Systems in Slaughterhouses Suitable for Developing Countries

Mahendra Kumar

2011

DAYA PUBLISHING HOUSE

DELHI-110 035

Published by	:	**Daya Publishing House** **A Division of** **Astral International Pvt. Ltd.** **– ISO 9001:2008 Certified Company** 4760-61/23, Ansari Road, Darya Ganj New Delhi-110 002 Ph. 011-43549197, 23278134 E-mail: info@astralint.com Website: www.astralint.com
Laser Typesetting	:	**Classic Computer Services** Delhi - 110 035
Printed at	:	**Chawla Offset Printers** Delhi - 110 052

PRINTED IN INDIA

ACKNOWLEDGEMENTS

This write-up was first prepared on behalf of the Food and Agriculture Organisation (FAO) of the United Nations, Rome around the year 1990 but could not be published for whatever may be the reason. The write-up has now been revised and updated.

The author acknowledges with thanks the help extended to him in various ways by the following personnel and organisations in the completion of this book.

Dr. S.K. Barat who initiated me into the area of animal byproducts while he was at Central Leather Research Institute (CLRI), Chennai and later he moved to FAO, Rome provided opportunities to work in a number of developing countries and visits to organisations in advanced economics, connected with animal byproducts.

My workplace, CLRI, which provided me full liberty to work on topics of interest to me.

Dr. Gunter Heinz, Senior Officer (Meat Technology), Animal Production and Health Division, FAO.

Dr. N. Muralidhara Rao, Dr. M.D. Ranganayaki, Dr. P.K. Sehgal, Dr. G. Biswas, Dr. C. Rose, Mr. T.S. Srinivasan, Dr. T.P. Sastry, Mr. K. Balasubramanian, Mr. T.S. Venugopal, Mr. K. Nagappan, Mr. N. Munuswamy and Mr. P. Paramasivam – my colleagues and co-workers in the Byproducts Area at the Institute.

My wife, Mrs. Kusum Agarwal for her cooperation and encouragement.

Mahendra Kumar

E-mail: mahendra_kumar@yahoo.com

vkagarwalji@gmail.com

FOREWORD

In any country, complete utilization of available resources is a key to sustainable economic development. The use of animals for food, fiber, and power has always been important in developing countries, but often the byproducts are not utilized because of lack of infrastructure or technology.

Among other things, a viable rendering industry can have a significant positive impact on the sustainability of animal agriculture and environmental quality for all citizens. The processing of otherwise low value organic matter from the livestock production, meat processing, food processing, and food service industries through rendering industry can drastically reduce the amount of waste. If not rendered, biological materials would be deposited in landfills, burned, buried, or inappropriately dumped with large amounts of carbon dioxide, ammonia, and other compounds polluting air and water. When rendered, these products are dried, stabilized, and recycled for animal feed, energy, fertilizer, and other uses. Many valuable materials from these animals can be collected before the rendering process further enhancing value.

In many parts of the world, rendering plants are large and expensive with sophisticated technologies designed for maximum throughput. However, it did not start that way. A century ago, most rendering plants were small, employing the simple operations of cooking and drying to extract the valuable components from the inedible portions of animal production. Even in its simplified versions, rendering done with limited inputs and minimal infrastructure can provide economic and environmental benefits.

This text will be most useful to those involved in animal agriculture in countries where byproducts are not fully utilized and the rendering industry is not yet developed. Byproducts and dead stock are often unused and a nuisance, causing environmental burdens. The practical science and technology presented in this text will improve knowledge leading to the improvement of the economic sustainability of animal rearing as well as providing benefits to animal and human health and the environment.

David L. Meeker

Ph.D., MBA

Senior Vice President, Scientific Services

National Renderers Association

Alexandria, VA, USA

PREFACE

Be it food, feed, fertilizer, manure, biofuel, pharmaceuticals, soaps, detergents, cosmetics, surgical products. life saving drugs, enzymes and so on, all kinds of products are made from animal byproducts.

The developing world owns around 3½ times as many cattle as the developed countries. After the animal is slaughtered, over 50 per cent of its weight is available as in-edible byproducts. The enormous quantities of such wastes as available in the developing world could, thereby, be well imagined. A better perspective can be had from the fact that in the United States of America alone, rendering industry is estimated worth about US$ 5 billion annually and regarded virtually important to animal agriculture and environmental sustainability.

Very unfortunate though, the developing world has not realized the economic, social and environmental potential of this vast resource. Very often, these are either allowed to be totally wasted or not put to optimum utilization. Economies apart, availability of raw materials in developing countries is invariably in small quantities. Appropriate technologies to process small throughputs in economically viable and environmentally clean manner had not been available. A number of books are now available for processing such small and limited quantities of raw materials, mostly published during the last 3 decades.

This write-up reviews and updates information available so far on the subject and also deals with such of those topics, not covered in earlier publications. With this exercise, one will find complete and comprehensive information on all aspects of byproducts utilization of small scale.

The author hopes that developing world will make use of this vast reservoir of information for the betterment of environment and upliftment of economic as well as social status of their people through the use of an invaluable raw material as available in their respective countries.

Mahendra Kumar

E-mail: mahendra_kumar@yahoo.com

vkagarwalji@gmail.com

Contents

Acknowledgements .... (*v*)

Foreword .... (*vii*)

Preface .... (*ix*)

List of Figures .... (*xix*)

List of Tables .... (*xxiii*)

1. **General Considerations**   **1**
   1.1   Definition 1
   1.2   Scope of the Technology 1
   1.3   Earlier Publications 2
   1.4   Source of Byproducts 2
   1.5   A Misleading Concept 3
   1.6   Impact of Non-utilization of Animal Offals 4
          1.6.1 Pollution of Environment 4
          1.6.2 Communication of Diseases 4
          1.6.3 Economic Loss 5
   1.7   Perpetual Wastage of Animal Byproducts –
          Possible Reasons 7
          1.7.1 Psychological Bias 7
          1.7.2 Lack of Awareness 7
          1.7.3 Lack of Technology 7
   1.8   Future Trends and Suggested Measures 8
          1.8.1 Use of Animal and Other Wastes 8

1.8.2 Strengthening of Animal Health Care
Measures 9
1.8.3 Raising of More Livestock 9
References... 10

2. **Types of Wastes Emanating from Slaughterhouses
and Current Practices of Their Collection and
Disposal in Developing Countries**   **12**
2.1 Availability 12
2.2 Collection and Disposal 12
2.2.1 Blood 13
2.2.2 Rumen Contents 13
2.2.3 Bone 13
2.2.4 Horns and Hoofs 14
2.2.5 Alimentary Tract 15
2.2.6 Gall Bladder 16
2.2.7 Soft and Fat Tissues 16
2.2.8 Keratinous Fibres 16
2.2.9 Poultry Offals 19
2.2.10 Effluents 20
References... 21

3. **Rendering**   **22**
3.1 Definition 22
3.2 Rendering Process.. 22
3.3 Products of Rendering 24
3.4 Aims of Rendering 24
3.5 Choice of Technology 25
3.6 Simple Cooking 26
3.7 Semi-moist Rendering... 29
3.7.1 Semi-moist Rendering vs Conventional
Rendering 29
3.7.2 Unique Features of the Technology 31
3.7.3 New Technology Package....................................... 31
3.7.3.1 Collection 33
3.7.3.2 Flaying 33
3.7.3.3 Entrails.. 35

3.7.3.4 Rendering 35

3.7.3.5 Mincing 36

3.7.3.6 Dry Meat Meal 36

3.7.3.7 Semi-moist Meat Meal 37

3.7.3.8 Use of Semi-moist Meat in Pig Feed 37

3.7.3.9 Pelleted Feeds 37

3.7.3.10 Stick Water 37

3.7.3.11 Bone 39

3.7.3.12 Compost and Horticulture 39

3.7.3.13 Treatment and Disposal of Effluents 40

3.7.3.14 Biogas Generation 42

Annexure-A: Semi-Moist Rendering Vessel:
Design and Operation 43

Annexure-B: Mincing 50

Annexure-C: Semi-moist Meat Meal and its
Water Activity 52

Annexure-D: Use of Semi-moist Meat Meal in
Pig Feed 55

References 56

4. Bone   **58**

4.1 Composition 58

4.2 Yield 58

4.3 Availability of Bones in Developing Countries 58

4.4 Uses of Bone 59

4.5 Handicrafts 60

4.5.1 Dense and Spongy Bones 61

4.5.2 Treatment 62

4.5.3 Preparation of Handicrafts 63

4.5.4 Quality of Finish 63

4.6 Bone meal 63

4.6.1 Bone Meal, Raw 63

4.6.2 Bone Meal as Livestock Feed Supplement 67

4.6.3 Bone Meal-steamed 68

4.7 Bone Ash 69

4.8 Uses 71
4.9 Other Bone Based Products 72
References 72

5. Blood   **73**
5.1 Introduction 73
5.2 Collection of Blood 73
    5.2.1 Collection on Rail 74
    5.2.2 Collection on Floor 78
5.3 Processing of Blood 80
    5.3.1 Blood Meal 80
    5.3.2 Lime Treated Blood 80
    5.3.3 Absorbed Blood 81
5.4 Sterilization 83
5.5 Nutritional Value of Blood 83
References 84

6. Rumen Contents   **86**
6.1 Definition 86
6.2 Rumen – A Living Factory 86
6.3 R.C. as a Useful Raw Material 86
6.4 Composition 87
6.5 Yield 88
6.6 Preparation of R.C. for Livestock Feed 88
    6.6.1 Sun Drying 88
    6.6.2 Ensiling 90
    6.6.3 Acid Preservation 93
6.7 Profile of Coarse and Fine Components of R.C. 93
6.8 Use 94
6.9 Relation Between Volume and Weight 94
6.10 Space for Sun Drying 95
6.11 Storage 95
References 95

7. Minor Products   **97**
7.1 Introduction 97
7.2 Glands and Organs 97
7.3 Intestines 99

7.4 Horns and Hoofs 100

7.5 Hair 100

7.6 Gall bladder 101

References 102

8. Sanitation and Hygiene   **103**

8.1 General 103

8.2 Building 103

8.3 Water 104

8.4 Transport Vehicles 104

8.5 Means of Communication 105

8.6 Dry Slaughtering 105

8.7 Plant Maintenance 106

8.8 Education Programme 106

8.9 Scheduled Diseases 107

8.10 Microbes 107

8.10.1 Shape 108

8.10.2 Size 109

8.10.3 Cell Structure 109

8.10.4 Motility 110

8.10.5 Multiplication and Growth of Bacteria 110

8.10.6 Spore Formation 113

8.10.7 Environmental Factors 113

8.10.8 Nutrition 113

8.10.9 Oxygen 114

8.10.10 pH (Hydrogen-ion Concentration) 114

8.10.11 Moisture 115

8.10.12 Temperature 115

8.10.13 Yeasts 116

8.10.14 Molds 116

8.10.15 Viruses 117

References 117

9. Biogas from Animal Wastes   **118**

9.1 General 118

9.2 Animal Wastes - An Ideal Source for
Biogas Production 119

9.3 Yield of Biogas from Different Types of
Raw Materials 119

9.4 Composition of Biogas 120

9.5 Optimum Parameters for Anaerobic Fermentation ... 120

9.6 Phases of Biogas Production 121

9.7 Biogas Plant 121

9.7.1 The Digester 123

9.7.2 Gas Holder 124

9.7.3 Inlet and Outlet Tanks 124

9.8 Products of Biogas Plant 125

9.8.1 Biogas 125

9.8.2 The Slurry 127

9.8.3 Solid Sludge 127

9.8.4 Liquid Sludge 128

9.9 Toxicity of the Slurry 129

9.10 Composting 129

References 130

10. Composting and Vermicomposting

10.1 Composting-Desirable Parameters 131

10.1.1 Porosity and Aeration 132

10.1.2 Moisture 132

10.1.3 Carbon, Nitrogen and Carbohydrates 132

10.1.4 pH 132

10.1.5 Atmospheric Temperature 132

10.2 Composting–Basic Dynamics 133

10.3 Microbes 135

10.3.1 Bacteria 135

10.3.2 Actinomycetes 136

10.3.3 Fungi 137

10.3.4 Protozoa 138

10.4 Materials for Composting 139

10.5 Methods of Composting 139

10.5.1 Pit Method of Composting 139

10.5.2 Stacking Method of Composting 140

10.5.3 Construction of Bunkers 140

10.5.4 Loading 142

10.5.5 First Turning 142

10.5.6 Second Turning 142

10.6 Stack Covered with Vegetable Matter 143

10.7 Windrow Method 143

10.7.1 Introduction 143

10.7.2 Culturing of Inocculant 144

10.7.3 Method of Windrow Composting 144

10.8 Benefits of Using Compost as Manure 147

10.9 Vermiculture/Vermicomposting 147

10.9.1 Introduction 147

10.9.2. Earthworms 148

10.9.3 Species of Earthworms Commonly
Used in Vermicomposting 151

10.9.4 Construction of Pit/Tank 153

10.9.5 Preparation of Vermicompost 154

10.9.6 Advantages of Using Vermicompost 155

10.10 Vermicomposting in Ground Heaps 155

10.11 Phosphorous Enriched Vermicompost 155

10.12 Vermiwash 156

10.13 Price 156

10.14 Composting vs Global Warming 157

References 158

11. Disposal of Condemned Materials   159

11.1 Introduction 159

11.2 Rules with Respect to Condemned Materials 159

11.2.1 Low Risk Materials 160

11.2.2 High Risk Materials 160

11.3 Destruction Methods 161

11.3.1 Incineration 162

11.3.2 Open Air Burning 162

11.3.3 Fixed Facility Burning 162

11.3.4 Air Circulation Incineration 162

11.3.5 Alkaline Hydrolysis 162

11.3.6 Anaerobic Digestion 163

11.3.7 Landfill 163

11.3.8 Burial 163

References 164

12. Treatment of Slaughterhouse Effluents **165**

12.1   Introduction 165

12.2   Pollutants 166

12.2.1 Biodegradable 166

12.2.2 Non-biodegradable 166

12.2.3 Biologically Accumulative 167

12.3   Sources of Wastes from Slaughterhouses 167

12.4   Nature of Slaughterhouse Effluents 169

12.5   Pollution Caused by Slaughterhouse Effluents 170

12.5.1 Biological Oxygen Demand (BOD) 171

12.5.2 Chemical Oxygen Demand (COD) 171

12.5.3 Nitrogen and Phosphorous 172

12.6   Microbes and Their Role in Effluent Treatment 172

12.6.1 Aerobic Decomposition 173

12.6.2 Anaerobic Decomposition 173

12.7   Aim of Effluent Treatment 175

12.8   Reduction of Effluents and Pollutants 176

12.8.1 Dry Slaughtering 176

12.8.2 Faeces, Blood, Injesta and Other Solids 177

12.9   Treatment and Disposal of Slaughterhouse Effluents177

12.9.1 Screening of Solids 178

12.9.2 Trapping of Grease 178

12.10   Soakage Pit 180

12.11   Septic Tank Followed with Subsurface Irrigation 181

12.12   Anaerobic Lagoon Followed by Waste
Stabilization Pond 182

12.13   Anaerobic Lagoon Followed by Aerated Lagoon 184

References 185

**13. Consolidated Statement of Products Possible to
Make from Slaughterhouse Wastes** **187**

References 194

Index **197**

LIST OF FIGURES

Fig. 2.1: From left to right: (1) Key chain, (2) Necklace, (3) Button   15

Fig. 2.2: (1) An external ear lobe from Zebu cattle; (2 & 3) Cut ears from Zebu cattle with ear lobes   17

Fig. 2.3: A bristle with split flags at the top end   18

Fig. 2.4: Shuttle cocks made from poultry feathers   19

Fig. 3.1: Meat being chopped manually   27

Fig. 3.2: Simple cooking of offals in a vessel   28

Fig. 3.3: Unit Operations for Fallen Carcass Utilization   32

Fig. 3.4: Carcass hoisted on four legs   34

Fig. 3.5: Carcass being placed on skinning bed   34

Fig. 3.6: Hoisted Carcass being flayed   35

Fig. 3.7: Drying of the Mince in a Pan into Meat Meal   36

Fig. 3.8: Dough being extruded in the form of pellets   38

Fig. 3.9: Dry pelleted feed   38

Fig. 3.10: Bunkers for Composting at BKT   40

Fig. 3.11: Sub-surface irrigation   41

Fig. Annex A-1: Photograph of Semi-moist Rendering Vessel   44

Fig. Annex A-2: Semi-moist Rendering Vessel–Sectional View   46

Fig. Annex A-3: Semi-moist Rendering
Vessel–Water and Steam in Circulation   47

Fig. Annex A-4: Photograph of a Brazier Suitable
for Coal as Fuel   48

Fig. Annex B-1: Meat Mincer in Operation   51

Fig. Annex C-1: Moisture v/s Water Activity   53

Fig. 4.1:   Dense and spongy parts of bone.   61

Fig. 4.2:   A pen holder with name printed (Buffalo bone).   64

Fig. 4.3:   An inlayed table   64

Fig. 4.4:   Inlayed figure of an elephant.   65

Fig. 4.5:   A candle stand (Camel bone).   65

Fig. 4.6:   A figure of a giraffe (Camel bone).   66

Fig. 4.7:   Jewellery articles (Camel bone).   66

Fig. 4.8:   A bone digester   68

Fig. 4.9:   Grill resting on brick walls, loaded with
bones and fired   70

Fig. 5.1:   Stunning with a sharp knife. An experienced
worker is able to reach the brain and puncture
the same in one single powerful stroke.   74

Fig. 5.2:   Carcass hoisted on rail and bled.   75

Fig. 5.3:   Carcass being bled into an oil drum.   76

Fig. 5.4:   An improvised hoist system.   77

Fig. 5.5:   A Manually operated hoist.   78

Fig. 5.6:   Blood collection on floor.   79

Fig. 6.1:   R.C. being dried in the open sun.   89

Fig. 6.2:   The mix to be ensiled, packed tightly inside
the drum double lined with polythene bags.   92

Fig. 8.1:   Cocci   109

Fig. 8.2:   Bacteria within capsules   110

Fig. 8.3:   Bacteria with various types of flagella   111

Fig. 8.4:   Bacterial growth curve at a constant temperature   112

Fig. 9.1: Chinese type of a biogas plant with a fixed dome gas holder. 122

Fig. 9.2: Cross section of a 6 m^3 Indian continuous type biogas plant. 123

Fig. 10.1: Plan of a Compost Bunker 141

Fig. 10.2: Perspective View of the Bunker 141

Fig. 10.3: Arrangement of pockets on the platform and Channel E. 145

Fig. 10.4: An Earthworm 149

Fig. 10.5: Kingfisher with an earthworm as its prey 150

Fig. 10.6: Clitellum around the worm's body 151

Fig. 10.7: Tanks for Vermicomposting 153

Fig. 12.1: Flow diagram illustrating the major pathways for the degradation of major categories of organic compounds under anaerobic condition. 174

Fig. 12.2: Screening and removal of solids from the effluents. 179

Fig. 12.3: A fat trap. 179

Fig. 12.4: Suggested effluents treatment and disposal system for very small slaughterhouse 180

Fig. 12.5: Suggested effluents treatment and disposal system for small slaughterhouse 181

Fig. 12.6: Subsurface irrigation trenches in parallel with trees planted along the bank. 182

Fig. 12.7: Suggested effluents treatment and disposal system for medium slaughterhouse (no mechanisation) 183

Fig. 12.8: Suggested treatment and disposal system for medium slaughterhouse (partly mechanised) 184

LIST OF TABLES

Table 1.1: Classification of edible, inedible and dual
purpose tissues from a slaughtered animal — 3

Table 1.2: Percentage weight of tissues in relation
to total live weight of the animal — 3

Table 3.1: Moisture of Rendered Tissues
Under 3 Categories of Processes — 29

Table 3.2: Moisture of Rendered Tissues Under
3 Categories of Processes on Moisture-
free Basis of the Solids — 30

Table Annex C-1: Moisture Content and Water Activity — 53

Table 5.1: Absorbent power of dry R.C. and wheat
bran and time required for drying of the mix — 82

Table 5.2: Essential Amino Acids in Beef and Cattle Blood — 84

Table 6.1: Analytical data of four types of R.C. — 87

Table 6.2: Analytical profile of fine and coarse
components of R.C. — 93

Table 7.1: Glands and Organs — 97

Table 7.2: Major Products and their Contents in
various Organs and Glands — 98

Table 7.3: Returns on End Products from Slaughter
of 1000 Animals — 98

Table 9.1: Yield of biogas from poultry, pig and
cattle manures — 120

(*xxiv*)

Table 10.1: Temperature and time of exposure required
        for destruction of some common pathogens
        and parasites in a compost stack/heap   134

Table 12.1: Sources of wastes in a slaughterhouse   168

Table 12.2: Composition of Effluents from different
        establishments   169

Table 12.3: Removal of organic matter and nutrients
        in a combined system   183

Table 13.1: Products from wastes of animals used for food   188

CHAPTER 1

GENERAL CONSIDERATIONS

1.1 Definition

The terms animal byproduct, animal waste or animal offal used in this monograph will specifically apply to inedible tissues of animal origin including fallen or dead stock; effluents also form part of the definition. Deviation, if any, will be stated so, as and when the subject is discussed.

1.2 Scope of the Technology

Generally speaking, except for big cities and towns, slaughter of the livestock in rest of the parts of countries in the developing world is invariably in small numbers resulting in the availability of limited quantities of raw materials fit for rendering. **The discussions here are exclusively for small quantities – say upto one tonne of rendering material inclusive of bone from all sources such as slaughterhouse, dead stock, butcher shop etc.**

The use of all types of animal based inedible tissues will form the subject of discussion in this monograph. Carcasses or parts thereof having been declared as unfit for human consumption, fallen carcasses, condemned offals and poultry wastes will thus all come under the category of inedible tissues and will be covered in the discussion.

Hides, skins and tanning, however, are not proposed to be discussed here. This is because of the fact that preservation of hides and skins and their conversion into leather is by itself an important commercial activity all over the world and vast literature is available on the subject.

1.3 Earlier Publications

At least seven publications have earlier appeared on the subject of utilization of animal byproducts in developing countries which being, FAO Agricultural Development paper No.75, *Processing and Utilization of Animal Byproducts*; FAO Agricultural services Bulletin No.77, *The Economics of Animal Byproducts*; FAO Agricultural services Bulletin No.79, *Handbook of Rural Technology for Processing of Animal Byproducts, Processing of Animal Byproducts in Developing Countries – A Manual, Recycling Process for Human Food and Animal Feed from Residues and Resources, Technologies for Value Realization of Carcass Byproducts in Developing Countries – A Handbook and Animal Byproducts Utilization through Semi-moist Rendering*. Yet another publication, *Sanitation and Hygiene in the Production of Rendered Animal Byproducts*, is of great use as it explains the role of microbs in rendering. The information will be of equal interest to meat industry as well, as the type of microbs both in rendering and meat industry being same. The present attempt could be considered as an extension of the earlier exercises. Consequently, it has been endeavoured to avoid repetition to the extent possible, provide additional information, and deal in greater detail such of those areas which have either not been treated well so far in the earlier publications or considered absolutely necessary for inclusion like semi-moist rendering, composting, vermicomposting, treatment and disposal of effluents, sanitation, hygiene and condemnation. Above publications have been quoted liberally particularly in respect of the information already published and one is advised to screen through these publications to get the full picture of the subject in its entirety.

1.4 Source of Byproducts

Animals are primarily slaughtered for food. However, not all the tissues of the animal are fit to be used as food; yet others may be considered as "dual purpose" tissues eg: they may or may not be used as food.

Broad classification of commonly edible, inedible and dual purpose tissues from slaughtered animals is given in Table 1.1.

The classification is dictated by a number of factors such as local or regional food habits, religion, customs and buying power of the consumer.[1, 2]

Table 1.1 : Classification of edible, inedible and dual purpose tissues from a slaughtered animal

Edible	Inedible	Dual Purpose
- Meat	- Horns	- Blood
- Brain	- Hoofs	- Oesophagus (gullet)
- Tongue	- Rumen Contents	- Intestines
- Sweet bread	- Hair	- Spleen
- Liver	- Feathers	- Udder
- Heart	- Condemned meat,	- Snouts
- Kidneys	offals, carcass	- Rectum
- Ox tail etc	or parts of it	- Lungs
	- Foetus etc	- Hide
		- Skin
		- Urinary bladder
		- Gall bladder
		- Bone
		- Stomach
		- Ears etc

1.5 A Misleading Concept

There is a prevailing impression in many quarters that there is no availability of animal byproducts in developing countries. This impression is not based on facts and needs to be clarified and corrected. An approximate percentage of different tissues on the total live weight of the animal as obtainable after its slaughter is given in Table 1.2.

Table 1.2 : Percentage weight of tissues in relation to total live weight of the animal

Tissue	Range%	Average%
1. Meat	35.0 - 40.0	37.5
2. Soft and fat tissues*	10.0 - 15.0	12.5
3. Hide & Skin	7.0 - 10.0	8.5
4. Bone	12.0 - 30.0	21.0
5. Rumen contents	10.0 - 15.0	12.5
6. Blood	3.5 - 5.5	4.5
7. Horns, Hoofs and tail hair	1.0 - 2.0	1.5

* Heart, brain, tongue, oesophagus, intestine, stomach, liver, lungs, kidney, etc.

On examination, one will conclude that items listed from 3 to 7 in the above table are invariably available as inedible offals in most of the developing countries; this is 48 per cent of the total live weight of the animal. This apart, some of the soft tissues, condemned meat, carcass or parts thereof are also available. Put together, it is evident that the availability of inedible wastes from slaughtered animals alone should not be less than 50 per cent of the total live weight of the slaughtered stock. In addition, fallen stock is one of the biggest resource in most of the developing countries. This is for the fact that proper health care measures of the animal are not available and there is a paucity of wholesome and balanced feed in these countries. In India, mortality rate is estimated to be above 10 per cent of livestock population annually; the availability of dead stock is reported to be over 41 million every year in the country. Hence, there is no paucity of animal wastes in the developing countries.

1.6 Impact of Non-utilization of Animal Offals

The non-utilization of animal offals creates a number of serious problems which are as follows:

1. Pollution of environment
2. Economic loss
3. Communication of diseases

1.6.1 Pollution of Environment

Chemically, the animal tissues are organic in nature and are amenable to fast degradation and putrefaction, if not suitably treated. This results into bad odours and pollution of environment in general. The surroundings of the slaughter place become breeding ground for mosquitoes, flies and vermin which, in turn, further pollute the environment and spread diseases.

It is not uncommon for the wastes being dumped and the effluents let out into streams, ponds, lakes and rivers in many developing countries. This results into ecological imbalances of the natural water sources causing destruction of the aquatic life and affecting their utility as well as scenic beauty.

1.6.2 Communication of Diseases

The importance of hygienic and safe disposal of condemned carcasses and meat which are unfit for human consumption is fully recognised as necessary for maintenance of public health by all civilized countries.[3] However, it is not uncommon in some of the developing countries that at times, the condemned meat is disposed off by spreading it out on the fields surrounding the slaughterhouse or by sending it to the city refuse dumps which will lead to parasitic diseases and infections to spread to livestock or human beings, flies to breed and bad odours to be caused.

Burying is recommended for disposal of a condemned carcass or small quantities of condemned meat, but if not carried out strictly according to scientific recommendations, this may set up areas of latent infection (eg. anthrax). The life cycle of certain helminthic parasites (eg. Taenia saginata, Taenia solium, Echinococcus granulosus, etc) will be perpetuated due to improper disposal of condemned meat, if meat infected with cysts of the parasites (eg. cysticercus bovis, cysticercus cellulosa, hydatid cyst, etc) are eaten by dogs.

1.6.3 Economic Loss

Under Chapter 13, a consolidated statement of various useful products which could be produced from animal wastes has been provided – a vast array of products to be useful in surgery, pharmaceuticals, photography, cosmetics, toiletries, leather, textiles, energy, food, feed and so on, could be made. Methods of production of a number of products such as meat meal, semi-moist meat meal, bone meal, blood meal, compost, vermi-compost, biogas, silage etc., have also been described under various other chapters.

The economy of most developing countries is agro based; as such, both feeds and fertilizers are essential and important inputs for livestock and agriculture respectively.

One of the main constraints that stands in the way of livestock, poultry, piggery and fishery development programmes in the developing countries is the lack of availability of adequate feeds with the requisite proteins, fats, minerals and vitamins at an economic price. The chronic shortage of balanced feeds is mainly responsible for lower productivity of animals, attendant with poor health, increased susceptibility to infection and disease, lower conception and birth rate, bearing of weak and dead offsprings, higher morbidity and mortality, lower production of milk, egg and meat.

Similarly, on the agricultural front, soils in most of the developing countries are deficient in phosphorous and often in calcium. Soils deficient in these minerals give lower crop yield and poor quality of grasses with higher fibre and lignin content. Animals grazing on grasses deficient in calcium & phosphorus are unable to make full use of the food available.

As already mentioned, protein meals, fats, minerals and organic fertilizers so badly needed both for livestock and soil, could be produced from animal wastes which unfortunately are allowed to be wasted in their entirety in almost all the developing countries. Regular application of bone meal and organic fertilizers corrects soil deficiencies and may help even in the conversion of unproductive soil into productive ones. This aspect has also been discussed in the chapter on "Composting and Vermi-composting". Similarly, the proteins, fats, minerals and vitamins needed for the production of balanced feeds for animal are available from products made from animal wastes. As the raw materials available indigenously are allowed to be wasted, feeds and fertilizers are to be imported at high costs. With the scarce foreign exchange position of the developing countries, adequate imports are not possible; this restricts the growth both of agriculture and animal industry.[4]

Against this background, the economic losses suffered by the country due to non-utilization of animal wastes are many fold, which are summarised below :

1. Direct wastage of a useful resource.
2. Pollution of the environment, ponds, lakes, streams and rivers.
3. Loss of aquatic life.
4. Loss of places of beauty and recreation.
5. Communication of diseases both to human beings and animal life causing morbidity and mortality and resultant loss due to these conditions.
6. Loss on account of non-establishment of industries possible to be based on raw materials of animal origin.
7. Loss of foreign exchange on importation of feeds, fertilizers and various other products.
8. Lack of quality feed to the animal industry.

9. Poor health of the animal.
10. Lack of quality organic fertilizers.
11. Poor quality of soil.
12. Poor crops.
13. Lack of availability of quality animal protein to humans.
14. Loss of employment opportunity.
15. And many more.

1.7 Perpetual Wastage of Animal Byproducts – Possible Reasons

There could be many factors responsible for the perpetual wastage of animal byproducts in the developing countries. Among many, the following are considered to be the most critical ones.

1. Psychological bias
2. Lack of awareness
3. Lack of technology

1.7.1 Psychological Bias

The term "byproduct" traditionally suffers from a kind of psychological bias. Byproducts are synonymous with subproducts meaning secondary products suggestive of lesser importance. Because of this psychology perhaps, no due importance has been given to byproducts utilization.

1.7.2 Lack of Awareness

There is a general lack of awareness of the economic potential of these resources and the possible contribution which their utilization can play in the country's economy. Endless number of products which are possible to be made have been listed in chapter 13. Psychological bias towards the byproducts is also responsible for the lack of awareness.

1.7.3 Lack of Technology

The lack of proper technology for the utilization of byproducts under the conditions of the developing countries is one of the most important factor responsible for their perpetual wastage. This is for the fact that slaughter of the animals generally takes place in small numbers spread over wide areas; the practice of centralised slaughter, as is the case in most of the developed countries, is most often absent. Consequently, the animal byproducts are available in small quantities and their processing by the technologies in vogue, is not economically viable.

The author has recently developed an approach of "Two Tier Technology" for processing of small quantities of animal byproducts.[1, 2, 5] Accordingly, the raw materials are conserved by the simple and cheap methods as a first step; the conserved materials could be pooled and later transported from many centres to a central processing facility. Preservation of the rendered meat, drying of blood by loading of the same on an organic absorbent, ensiling of rumen contents as explained under the chapters on "Rendering", "Blood" and "Rumen Contents" respectively are some of the examples of simple and cheap conservation methods.

Additionally, a new technology "Semi-moist Rendering" has been recently reported in detail in a book[6], which is ideally, suited for making use of raw materials available in small quantities.

1.8 Future Trends and Suggested Measures

The human population in all the developing countries is on the increase. Moreover, food habits are also changing towards consumption of more meat. As a result, many countries which had agricultural surplus for export during the past, have become net importers both of agricultural produce as well as meat; Nigeria is a classical example of this. Unless timely measures are taken to produce more meat in these countries, the problem may assume unmanageable proportions.

The following three measures are considered necessary to improve the situation.

1. Use of animal and other wastes.
2. Strengthening of animal health care measures.
3. Raising of more livestock.

1.8.1 Use of Animal and Other Wastes

Animal wastes apart, many kinds of other wastes such as agricultural residues, left overs from fruit, vegetable, fish and meat markets, hotel residues and left over foods, certain types of industrial wastes (molasses, yeast, bran, rice polish and so on) etc are available in most of agro based economies. All these materials should be salvaged and converted into balanced feeds and fertilizers and made available at reasonable price.

1.8.2 Strengthening of Animal Health Care Measures

Strengthening of health care measures through improvement of veterinary services is an unavoidable component for increasing meat production even from the existing stock of livestock. In the absence of proper health care, the stock suffers from innumerable diseases as discussed under 1.6.2. This causes poor health, morbidity and mortality resulting into lower meat production.

1.8.3 Raising of More Livestock

Obviously it is possible to register significant increase in meat production from the existing stock itself by converting the agricultural and other wastes into balanced feeds and strengthening of animal health care services. This potential could be judged by the fact that developing countries with roughly 3½ times as many cattle as the developed countries and 5½ times that of U.S.S.R and Eastern Europe produce only about half the meat of the former and just one third more of the latter. With better inputs in terms of quality feeds and health care, doubling of the meat production with the present stock itself may not be considered beyond reach.

Besides, over 60 per cent of the permanent pasture of the world is in the developing region. The vast central low lands of South America, most of the Central and East Africa, large areas of arid and semi-arid lands in Near and Middle East and high land regions of South East Asian countries offer immense possibilities for livestock production. The extensive grassland areas of Mato Grosso and Sudem in Brazil including the Pantenal region offer considerable potentiality for livestock development. The River Plate countries of Argentina, Paraguay and Uruguay also have substantial possibilities along with Mexico. Among the Andean Countries, Colombia in particular offers the maximum scope. The extensive grazing areas of Ethiopia and the savanna region of Sudan afford first class livestock raising potentiality still to be explored. The productive capacity of the rangeland in the lower and upper Juba region of Somalia could be raised 2–3 times as at present. Nigeria, Mali, Mauretania, Botswana, Swaziland, Lesotho and Malagasy have similar potential resources awaiting exploitation. Moreover, eradication of tsetse fly would open up 7–9 m.km^2 of first class grazing land in Africa. In Asia, the Near East, Thailand, Indonesia and the Philippines offer similar scope of development.

Developing countries need to accord appropriate priority to effective commercialization of their animal byproduct resources. Lack of awareness and still worse, misconception about these resources at the decision making level has often prevented or retarded such development. The best way to tackle this situation is to establish and demonstrate the economic viability of these materials through pilot scale production and marketing.

A workable strategy for development of byproduct resources in a Developing Country must cover its basic needs, economic conditions and realities of market. Innovative but low capital intensive technology developed through indigenous R & D is the best suited to Developing Country conditions rather than the so called industrial production model of the West based on wholly imported inputs. Vertical integration of abattoir in Mexico, China and Mongolia in the public sector and in Costa Rica in the cooperative sector, horizontally integrated private sector collection and processing of animal byproducts in Brazil and the utilization pattern of fallen animals in India provide some of the examples or working models to choose from for further adaptation and improvement.[7]

References

1. Mahendra Kumar (1987) : *Processing of animal byproducts in developing countries – a manual.* Commonwealth Science Council, Marlborough House, Pall Mall, London Sw1Y 5HX UK.

2. Mahendra Kumar (1989) : *Handbook of rural technology for processing of animal byproducts (FAO Agricultural Services Bulletin No.79).* Food and Agriculture Organisation of the United Nations, Rome.

3. Don A. Franco (1977): *Sanitation and hygiene in the production of rendered animal by-products.* Published under the auspices of the Animal Producers Industry, The Fats and Proteins Research Foundation and The National Renders Association (USA).

4. Mann, I (1962) : *Processing and utilization of animal byproducts (FAO Agricultural Development Paper No. 75)*. Food and Agriculture Organisation of the United Nations, Rome.

5. Mahendra Kumar (2001): *"Two-tier technology – an efficient tool for carcass byproducts utilization in developing countries"* in "Technologies for value realization of carcass byproducts in developing countries – a handbook". Ramasami, T. Ranganayaki, M.D. and Rajagopal, N.R. Central Leather Research Institute, Chennai-600020 (India).

6. Mahendra Kumar (2007): *Animal byproducts utilization through semi-moist rendering*. Daya Publishing House, 1123/74, Deva Ram Park, Tri Nagar, Delhi-110035 (India).

7. Barat, S.K (1989) : *Animal byproducts in the developing world – an overview :* paper presented at the workshop on animal byproducts utilization for developing countries jointly organised by CSC, London and CSIR, New Delhi between March 6-20, 1989 at the Central Leather Research Institute, Chennai-600020 (India).

CHAPTER 2

TYPES OF WASTES EMANATING FROM SLAUGHTERHOUSES AND CURRENT PRACTICES OF THEIR COLLECTION AND DISPOSAL IN DEVELOPING COUNTRIES

2.1 Availability

A broad list of various kinds of wastes or rather byproducts as may be available from slaughtered animals in the developing countries is given below :

Blood, rumen contents, bones, horns, hoofs, urinary bladder, gall bladder, intestines, uterus, rectum, udder, foetus, snout, ear, penis, meat trimmings, hide and skin trimmings, condemned meat, condemned carcass or parts thereof, oesophagus (gullet), hair of various kinds (like cattle tail hair, cattle ear lobe hair, pig bristle and pig body hair), poultry offals (feathers, blood, intestinal tract and head) and effluents.

The list provides a comprehensive picture of the various types of wastes available but it does not necessarily mean that all of them are available in every place. Many of the dual purpose tissues as discussed in Chapter 1, are consumed as human food but still there are wide differences from country to country and even from region to region in the same country. Generally speaking, the number of wastes as available in most of the places is much less than provided in the list.

Hides and skins are now either exported or converted into leather in many developing countries. Hence hides and skins are now considered as co-product (of meat) and not byproduct or waste. Many developing countries have very large and modern tanning industry – India is one such example.

2.2 Collection and Disposal

An overview of the prevailing practices of collection, processing and disposal of various wastes as available in the developing countries is offered. This will enable the reader to visualise the colossal waste taking place at present and at the same time an opportunity the situation offers for their proper and economic utilization.

2.2.1 Blood

Generally speaking, blood is allowed to be wasted in most of the developing countries. After slaughter, the blood may be allowed to run. The same coagulates and blocks the drains where it putrefies and becomes a breeding ground for flies and insects. Once putrefied, it emits very bad odours polluting the whole environment.

In some countries like Philippines and Southern Nigeria, blood is collected for human consumption; in some others, it may be converted into blood meal, though on a limited scale. The methods of collection, its processing into food and blood meal etc. according to local practices have all been described in detail.[1, 2]

The use of blood on scientific lines has been fully described in the chapter on "Blood".

2.2.2 Rumen Contents

Like blood, rumen contents are also allowed to be wasted in most of the developing countries; the rumen is emptied of the contents which, in turn, may be dumped in the vicinity of the slaughter place itself causing offensive odours and providing a breeding ground for mosquitoes and flies etc.

In some places, rumen contents are composted for use as manure. However, rumen contents can find more remunerative uses – production of biogas and livestock feed are two examples. These aspects have been discussed in detail under the chapters "Biogas from Animal Wastes" and "Rumen contents" respectively. A complete chapter is also devoted on composting and vermicomposting.

2.2.3 Bone

The availability of raw or green bones from slaughtered animals as an inedible material is almost negligible in the developing countries. Normally the meat vendors buy dressed carcass as a whole, or as halves or quarters and sell the meat into smaller cuts along with the bones to the consumer. The consumer, in turn, cooks the meat along with the bones. Big and long bones like femur, tibia etc., are at times available from meat vendor shops. These too, however, may be chopped and sold as soup stock.

In some of the countries, the skull may be deboned at the slaughterhouse itself. In such cases, skull and jaw bones may be available as green bones. Other bones are rarely available at the slaughterhouse.

While bone is also allowed to go as a complete waste in many countries, in certain others, these are collected and utilized.

In countries like India, Pakistan, Bangladesh and Thailand, collection of bone is well organised. In India, there is also a fairly good bone glue, DCP (Dicalcium phosphate), bone meal, bone char, ossein, edible gelatine, pharmaceutical gelatine and photographic gelatine industry. Sinews obtained during crushing of bone are also utilized for the manufacture of glue. India, Pakistan and Bangladesh also have sizeable export of crushed bones. In addition, India also exports ossein which is obtained by the process of demineralisation of bone and is a raw material for the manufacture of high quality gelatine such as edible, pharmaceutical and photographic.

Processing of bone into fancy articles and bone ash etc, has been described in the chapter on "Bone".

2.2.4 Horns and Hoofs

The story about horns and hoofs is not very dissimilar than that with many other offals - most often horns and hoofs are also allowed to be wasted.

Horns, along with horn pith, are cut from the skull of the animal and thrown away as waste in many countries. Because of the bacterial decay, the horn loosens after some time, and gets separated from the pith.

In many countries, the lower leg (Shank) along with the hoof may be sold as a soup stock; the same is treated in a boiling water bath as to loosen hair and hoofs. Thereafter, the hair is scrapped off and hoof is tapped out by hitting the leg against a hard surface. The hoof may be thrown away in dust bin as a waste.

In countries like India, Pakistan and Bangladesh, there is a well organised collection of bone, horns and hoofs. A small quantity of these are used in handicrafts while bulk is converted into raw or steamed horn & hoof meal. Many countries like Japan, import both horn and hoof meal as well as raw horns and hoofs.

A variety of handicrafts like key chains, buttons, fancy and jewellery articles are made in many countries as shown in Fig. 2.1.

Fig. 2.1 : From left to right : (1) Key chain, (2) Necklace, (3) Button

More discussion is available on horns and hoofs in the chapter on "Minor Products".

2.2.5 *Alimentary Tract*

Almost all the parts of the alimentary tract like oesophagus (gullet), stomach, intestines and rectum are generally consumed as food in most of the developing countries. Organs like stomach, intestines and rectum are emptied of their contents and thoroughly washed before these are used as food. Soup made out of intestine is considered as a delicacy in some of the countries.

The use of intestine as food is not so common in countries like India, Pakistan and Turkey where these are processed into sausage casings. In all these three countries, casings are also used for the production of surgical catgut; cattle as well as sheep and goat intestines could be used for the manufacture of surgical catgut, sports guts and musical guts. More information on intestine is available in the chapter on "Minor Products".

2.2.6 Gall Bladder

Gall bladder is treated as a total waste in almost all the developing countries. The same is severed with the help of a knife and thrown away as waste. Gall bladder contains a dark yellow to green syrupy liquid called "bile" which is bitter in taste.

At least in two countries viz. Philippines and Sudan, bile is consumed for edible purpose. In northern parts of Philippines, meat is cooked along with requisite quantity of bile for the preparation of a popular dish called "Pinapaithan"; bile imparts a characteristic taste to the meat which is liked by the people. In Sudan, bile is sprinkled on raw meat cut into small pieces; this dish is used as a starter before the main course and is attributed to invoke appetite.

More information on bile is available in the chapter on "Minor Products".

2.2.7 Soft and Fat Tissues

Various soft tissues like lungs, kidney, fat tissues, urinary bladder and meat trimmings are utilized as food in most of the developing countries. These tissues are normally available at relatively less prices compared to the meat and are generally consumed by poor people.

2.2.8 Keratinous Fibres

A variety of keratinous fibres like cattle tail hair, cattle ear lobe hair, horse tail and mane hair, pig body hair, pig bristle etc. are available as a result of the slaughter of the animals. These materials are very precious and find use in the manufacture of different kinds of brushes – wall painting, artist, shaving and many more kinds. While this is so, this important raw material also does not find any use in many developing countries; in some others, no doubts, hair is suitably utilized.

In many countries of Africa, the hairy tail is cut and thrown away as waste; in some others, the same may be dried in the sun, particularly when the hair is long and bushy, and used as a duster.

In India, cattle tail hair is used for the manufacture of cheap varieties of wall painting brushes[1]; other types of hairs like horse tail and horse mane hair could also be used similarly. Long and strong horse tail hair is also used as violin bow hair.

Cattle ear lobe hair is a very precious hair as the same is used for the manufacture of finest and most expensive artist brushes. This is the hair which grows at the rim of the external ear lobe of cattle; the best quality hair being available from large sized zebu cattle (Fig. 2.2).

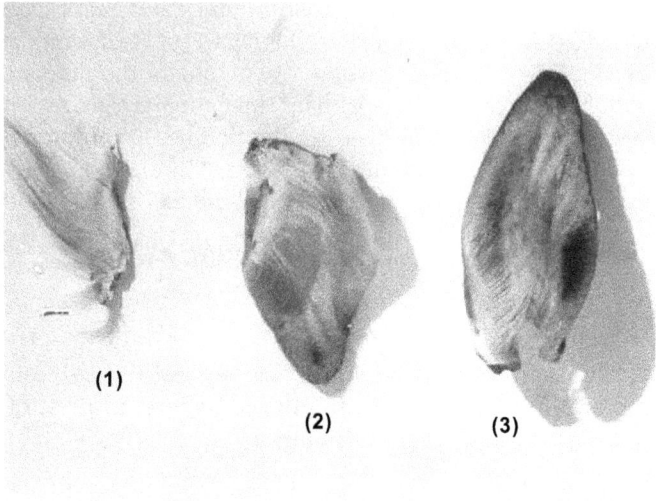

**Fig. 2.2 : (1) An external ear lobe from Zebu cattle;
(2 & 3) Cut ears from Zebu cattle with ear lobes**

Zebu is the most popular breed of cattle raised in almost all the developing countries and hence offers a great potential for the availability of this special kind of hair. Unfortunately, however, this resource is almost entirely wasted in practically all the developing countries. In a few countries of Africa like Botswana, Kenya and Zimbabwe where slaughterhouses have been established for export of meat essentially to Europe, the ear lobe is conserved and exported.

In India and China, bristle and pig body hair are used for the manufacture of brushes of various kinds[2]. Bristle is the term used for the hair exclusively growing along the backbone and neck of the pigs, hogs and wild bores. Bristle is long, stiff and wiry, which is thick at the root and gradually tapers down in thickness towards the tip (Fig. 2.3). The tip end bifurcates into many flags, which helps in the retention of more paint than a single strand, resulting into finer and even spread of the paint on the surface to be painted.

In India, the entire brush industry is in the cottage sector providing employment to a large number of people. Unfortunately, as has already been discussed, all types of hair available in most of the developing countries are wasted but they have a chance to learn from Indian and Chinese experience and develop brush industry which admirably suits to the conditions of developing countries.

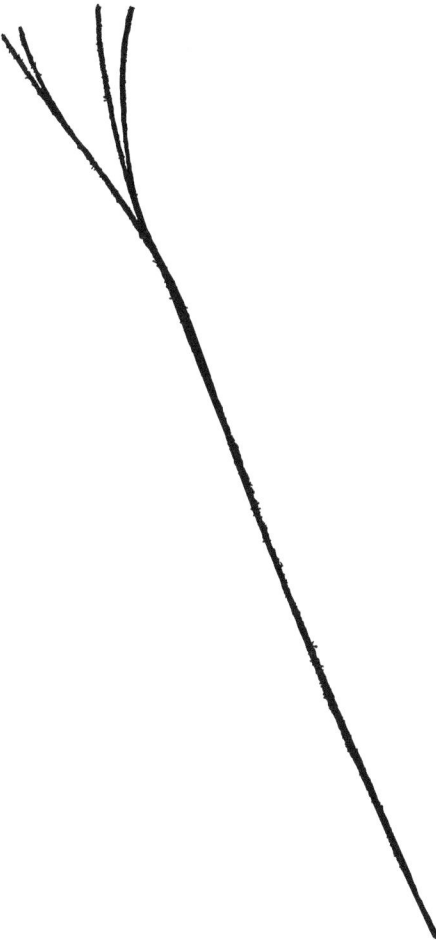

Fig. 2.3 : A bristle with split flags at the top end

Collection and conservation of different types of hair is rather simple and has been discussed in detail in the literature.[1,2]

2.2.9 Poultry Offals

Generally speaking, poultry offals of all description are a total waste in almost all the developing countries with only a few exceptions and that too on limited scale.

In Manila (Philippines), a commercial plant makes use of all the offals made available from a large poultry slaughterhouse, to process into poultry offal meal. This has become possible because of the availability of large quantities of offals and that too from a single source. In most of the developing countries, including the rest of Philippines, size of slaughter at a particular given time and place is small and hence the offals get perpetually wasted. The approach of "Two Tier Technology" as discussed in Chapter I, can be useful in this situation.

In India and Pakistan, selected poultry feathers (15 cm and above) are used for the manufacture of badminton shuttle cocks (Fig. 2.4). Like brush, shuttle cock manufacture is done on cottage scale both in India and Pakistan, providing employment to a large number of people. Other countries can take advantage of this technology.

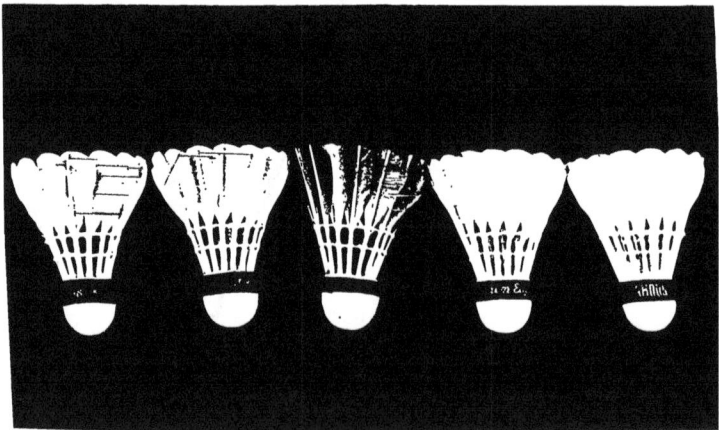

Fig. 2.4 : Shuttle cocks made from poultry feathers

Details of collection, conservation and processing of poultry feathers into animal feed and shuttle cock are available in the literature.[2]

In India, poultry feathers and human hair are also used for the manufacture of lysine–an amino acid which finds use in food and feeds. The process consists in hydrolyzing the hair or feathers by acid hydrolysis and then isolation of lysine from the hydrolysate.

Poultry feather hydrolysate has also been found useful as leather filler[3] as well as in leather finishing.

2.2.10 Effluents

Slaughterhouse effluents is one of the most serious causes of environmental pollution, bad odours and health hazards in almost all the developing countries; the treatment and disposal of effluents being almost non-existent. The problem is more serious in the cities than the rural side because of larger number of animal being slaughtered in the cities; sufficient land is also not available in the cities for the natural decay and degradation of the wastes as well as that of effluents.

The city sewage systems are not either efficient or not geared to take the load of slaughterhouse effluents. This results in the blockage of the drains etc. in the slaughterhouse with all the attendant consequences. There are no sewage systems in the rural side and hence the question of disposal through such a system can not be thought of. Many a times, the effluents are let out into environment. In some places, soakage pits have been constructed.

Modern systems of treatment have been established in some of the countries, though on limited scale, but the results have been disappointing. The maintenance has been poor resulting even in the closure of the plants in some places.

Biological treatments have been tried in Philippines and India with reasonably good success, though on limited scale. Basically, these systems are based on anaerobic lagoons followed with disposal either into aerobic ponds or soakage pits. At Maya Farms in Manila, effluents and wastes of 15,000 pigs and the effluents from the slaughter of a small number of pigs at the farm slaughterhouse, are successfully treated by the biological methods; a brief discussion of this is available in the chapter on "Biogas from

Animal Wastes". One is advised to refer to the literature for more details.[4]

References

1. Mahendra Kumar (1989) : *Handbook of rural technology for the processing of animal byproducts (FAO Agricultural Services Bulletin No.79).* Food and Agriculture Organisation of the United Nations, Rome.

2. Mahendra Kumar (1987) : *Processing of animal byproducts in developing countries – a manual.* Commonwealth Science Council, Marlborough House, Pall Mall, London SW1Y 5HX UK.

3. Sastry, T.P., Sehgal, P.K., Gupta, K.B. and Mahendra Kumar (1986): Solubilized keratins as a novel filler in the retanning of upper leathers. *Leather Science,* Vol. 33 (12).

4. Felix D Maramba, Sr. : *Biogas and waste recycling – the Philippines experience.* Maya farms Division, Liberty Flour Mills, Inc. Metro Manilla, Philippines.

CHAPTER 3

RENDERING

3.1 Definition

The word "render" means "to melt" or "to clarify" by melting. Accordingly, originally the term "rendering" meant melting out the fat from the fat tissues of animal origin with the application of heat and its recovery from the same. With the development of the animal byproducts industry over the time, the term today stands to mean the production of whole gamut of products of rendering which being meat meal, meat-cum-bone meal, semi-moist meat meal, bone meal and fat from the animal tissues. Evidently, the processing of blood, horns, hoofs and poultry feathers into products like blood meal, horn and hoof meal and poultry feather meal, where no fat is recovered, falls outside the purview of rendering.

3.2 Rendering Process

For centuries, after the animal was slaughtered, meat, hide and skin alone were put to use; the rest of the carcass was considered as "waste".

In the early 1890's, some select meat residues from lard rendering were used for chicken feed. Around 1900, poultry feeders on the Pacific Coast, California, USA started using dry tankage (solids from rendering process) in poultry rations. Around the same time, this new feed was also used in hog rations. Thus was born a new era in the utilization of rendered tissues in livestock feeding.[1]

Initially, rendering was usually done by cooking the tissues alongwith added water for long periods in an open vessel or pan; the process being called "Open Pan Rendering."[2,3] Subsequently, another two improved versions of open pan rendering were developed christened as "Simple Cooking"[2,3,4] and "Kettle Rendering."[3] Out of these 3 processes, open pan rendering and kettle rendering are not any more worthy of use, even though investment on both of them is very low, they end up with low quality products and pollute the environment. Simple cooking may however, still have some validity and this aspect will be discussed elsewhere.

After the above mentioned simple processes, more mechanised systems started appearing; the first in the order being "Wet Rendering."[2, 3, 4, 5, 6, 7] Unlike the earlier processes where rendering is carried out at atmospheric pressure, wet rendering is carried out at pressure under steam in a pressure vessel. Around the middle of 20th century, a more mechanised system of rendering was developed known as "Dry Rendering."[2, 3, 4, 5, 6, 7] This not only almost replaced the earlier processes but remained in use for quite long period and infact till today in many places. It may be desirable to mention that both wet rendering and dry rendering are batch processes.

Thereafter, rendering technology has seen phenomenal progress over the years. Highly sophisticated automatic plants such as Duke Continuous System, Atlas Low Temperature Wet Rendering System, Pfaudler Conversion System, Centri Meal System etc. are available nowadays. In some systems, rendering is carried out at low temperature (80°C or so) which is basically different than the conventional methods where high temperatures are followed. In terms of technology, these could be considered similar to chemical plants requiring highly trained personnel for their running and maintenance. Such systems are economical only for large through-puts and need large investments. Such inputs are not available in most of the developing countries and thereby cannot be recommended for their introduction.

Based on the R&D and field work of over 3 decades, this author has developed a new process of rendering based on new concept called "Semi-moist Rendering." It is a zero waste technology, simple in operation, economically viable, environmentally clean, eco-friendly, rendered products being of highest possible quality and yet requiring limited investment on infrastructure and equipment. The process will be discussed in some detail elsewhere but for complete detail, one is advised to refer the original publication.[8]

The process will be most suitable for adoption in developing countries.[8,9]

3.3 Products of Rendering

Depending upon the raw material used, there are four products obtainable during conventional rendering which being meat meal, meat-cum-bone meal, bone meal and tallow.

Semi-moist rendering as mentioned above, is also a rendering process but its rendered meal is called "Semi-moist Meat Meal" and not meat meal.

Meat meal does not mean a product made out of meat or soft tissues alone; it may partly contain bone as well. A meal having its protein content at 55 per cent and above and phosphorous upto 4.4 per cent (equivalent to 10 per cent of phosphoric acids) on moisture free basis is normally called as meat meal in the trade. The protein, however, should have been contributed by meat offals and bone and not by horns, hoofs, hair, blood, rumen contents or any other matter.[2,3,10] A meal made out of meat offals alone may contain upto 80 per cent crude protein content. Meat meal is also called by many other names like digester tankage, meat scrap and dry rendering tankage.[4]

Meat-cum-bone meal is a product containing crude protein content below 55 per cent and phosphorus more than 4.5 per cent on moisture free basis.[4, 10] More bones are present in this product than meat meal. The product is also known as meat and bone scrap.[4]

Bone meal is obtained when only bones have been rendered.

Semi-moist meat meal has a protein content of around 65 per cent, fat about 5 to 7 per cent with phosphorous being almost negligible on moisture free basis.

Tallow is the fat obtained during rendering process. When obtained from beef cattle, it is known as beef tallow; in the cases of sheep and goat, it is called mutton tallow. The swine fat, however, is known as lard.

3.4 Aims of Rendering

There are three main aims of rendering which are as follows :

1. Separation of fat from the tissues rendered.

2. Elimination of moisture from the fat free tissues or meal.

3. Sterilization of the rendered products.

The fat has to be separated for two reasons. It fetches a higher price compared to the meat meal. This apart, meals having higher percentage of fat have a poor keeping quality. The fat present in the meal gets oxidised and becomes rancid over the time. Rancidity is not desirable as it affects palatability of the meal.[2, 3, 4] A meal having fat content upto 10 per cent has a reasonable storage life[2,3,4]; lower the fat content better it is for storage.

Elimination of moisture or water from the dry meat meal is rather essential. Meat offals contain around 75 per cent water and its elimination reduces the weight and bulk of the material. Moreover, excess water helps bacterial growth. Dry meat meal should not have more than 8 per cent moisture content.[2,3,4]

As against the conventional dry meat meal, the semi-moist meat meal is preserved wet by controling its water activity, very much in the same fashion as the pickles are preserved; the water content of the meal ranging between around 40-45 per cent.

The rendered products, viz. the meals as well as the fat are used in livestock rations. As such, they should ideally be sterile. The temperature at which the tissues are rendered help sterilization of the rendered products. Subsequent handling procedures should also be such that the products do not get contaminated with bacteria particularly with pathogens.

3.5 Choice of Technology

The choice of a process to be selected depends upon the ground realities such as quantum of raw material available, type of raw material, investment available, capacity to absorb the technology and many more.

With his background of having extensively worked in the developing world, the author feels, that except in situations where meat is exported like Zambia, India, Kenya and so on and perhaps big cities slaughtering large number of animals, highly mechanised rendering systems are a total failure in developing countries. Nigeria is one such example. Municipal abattor in Mumbai (India) met with a similar fate.

In the scenario as discussed above, one is perhaps left with the choice of three processes to be recommended for introduction in the developing world which being

1. Dry Rendering
2. Simple Cooking and
3. Semi-moist Rendering

Dry Rendering can be considered for big cities where large number of animals are slaughtered. Even so, before taking a decision, it should be ensured that enough raw material is available for rendering for the size of the plant under consideration. Raw material for rendering consists of the inedible tissues available from the slaughter of the animal, as listed in Table 1.1 in Chapter 1. Except blood (mostly in Muslim countries), and bone in some countries, most of the so called "inedibles" are consumed as food in the developing countries. Should enough raw materials be not available, other alternatives will have to be considered for the type and quantities of the raw materials available. The picture in the Latin American and Caribbean countries, however, is different where many of the inedible tissues will be available for rendering.

As the name implies, simple cooking is a very simple process and can be useful in the remotest places for processing and making use of raw materials even while available in very limited quantities. Such conditions do exist in many developing regions and so worthy of discussion.

Semi-moist Rendering will be the process of choice in the country side of the developing countries where invariably limited number of animals are slaughtered. The process has considerable flexibility in as much as the same can be suitably tailored to process from rather modest quantities to something upto one tonne of raw material.

Details of Dry Rendering are not proposed to be discussed here. One can get full processing details from the literature.[8] Simple cooking as well as semi-moist rendering will be discussed in some detail. However, for comprehensive information of semi-moist rendering, one is advised to go to the publication on the subject.[8]

3.6 Simple Cooking

Simple cooking can not strictly be called rendering as there is no attempt of separation of fat from rendered tissues in this process, as will be seen from the description here under. The process, however, merits mention because it is one way of making use of the animal offals particularly when available in small quantities.

Simple cooking of the animal offals may be followed only in those places where the cooked mass could be fed directly to the livestock without its being converted into a dry product. For this reason, the cooked mass should be fed to the animal soon after its preparation, as otherwise, it may get contaminated with bacteria and become unsafe for feeding. While this is very simple and cheap process for the disposal of offals, particularly when available in small quantities, it could be successful only if there is a very good coordination between the producer and the pig farm. The process details are as follows :

1. All the available tissues are chopped into smaller pieces (2-3 cm^3). Hand mincing of small quantities of meat offals is possible over a wooden log with the help of a knife, as shown in Fig.3.1.

Fig.3.1 : Meat being chopped manually

The chopped mass is loaded in a vessel and water sufficient to cover the loaded mass is added (Fig. 3.2). The vessel may be covered with a lid during the period of cooking.

Fig. 3.2 : Simple cooking of offals in a vessel

2. Depending upon the size of the material, the contents are boiled continuously for about 1 hr till cooked well.

3. Other ingredients required to balance the feed like wheat bran, rice bran, oil cake, rumen contents, cassava flour etc., are added and mixed thoroughly with the offals already cooked. The mixing of the ingredients is so adjusted that the meat offals in the total mix is at around 10 per cent. The mix is cooked again for about 1/2 hour and brought to a porridge like consistency.

The product, as prepared, is ready and could be fed to the pigs. The product is rich in proteins, carbohydrates, fats and vitamins. Minerals could also be added to make it a completely balanced diet.

3.7 Semi-moist Rendering

Semi-moist Rendering is a new technology of rendering which has been developed by the author after extensive R&D and field work over a period of around 3 decades. In its present form, the method is most suited for processing smaller quantities of raw materials either from slaughtered animals or dead stock. Invariably, at a particular given time and place, only limited quantities of raw material are available in most of the developing countries. Consequently, the process is most suited under the conditions in the developing world. Some details of the process are reported; full details are available in the publication on the subject.[8]

3.7.1 Semi-moist Rendering vs Conventional Rendering

The classification of rendering processes can be done on the basis of moisture content of tissues, soon after they ensue from rendering; consequently, all the processes in vogue will fall into two categories viz; dry rendering and wet rendering. This author has developed a new process of rendering to be called "Semi-moist Rendering" wherein the moisture of the tissues as they ensue after rendering, falls in between the earlier two kinds of processes of dry rendering and wet rendering. New process apart, the author also developed the whole package of technology, which covers all aspects associated with rendering including treatment of effluents and their utilization etc.

With a variation of about minus 5 per cent, the moisture of the rendered tissues under 3 categories of processes is shown in Table 3.1.

Table 3.1 : Moisture of Rendered Tissues
Under 3 Categories of Processes

Type of Process	Percentage of Moisture of Tissues as they ensue soon after Rendering	Percentage of Solids
Wet Rendering	80	20
Semi-moist Rendering	55	45
Dry Rendering	08	92

Translating the data in Table 3.1 on moisture free basis of the solids, the percentage of moisture held by solids in the three respective processes is shown in Table 3.2.

Table 3.2 : Moisture of Rendered Tissues Under 3 Categories of Processes on Moisture-free Basis of the Solids

Type of Process	Quantity of Tissues as they ensue soon after Rendering	Percentage of Solids held by the Rendered Tissues	Percentage of Moisture held by the Solids on Moisture Free Basis
Wet Rendering	100 weight units	20	400.00
Semi-moist Rendering	100 weight units	45	122.20
Dry Rendering	100 weight units	92	8.70

Rendered tissues can be stored safely for reasonably long time at a moisture content of less than 10 per cent. As such, no further drying is necessary for storage of dry rendered produce, known as meat meal. However in the case of wet rendered material, 391.3 (400-8.7) per cent moisture will have to be expelled out to obtain 100 weight units of meat of the same as to be at par in moisture content with dry rendered one; as opposed to this, only 113.5 (122.2-8.7) per cent moisture will be required to be driven out in the case of semi-moist rendered stuff.[8]

In other words, almost 3.5 times (391.3-/113.5 = 3.44) moisture is required to be driven out of the wet rendered material as compared to semi-moist rendered one; this will mean considerable savings on energy, land and infrastructure costs in the case of semi-moist rendering option. This basic difference becomes the basis for the new process to called **"SEMI-MOIST RENDERING"**. In the case of semi-moist rendering, the costs are further slashed down, because the rendered tissues are to be preserved wet as such, by controlling the water activity.

Considering of costs further, it will be relevant to mention that the reduction of moisture of the meat meal in the case of dry rendering, has not been without the tag of cost addition. Steam has been used to drive out moisture from the rendered tissues as to arrive at a moisture content of around 8 per cent.

Hence cost of production in semi-moist rendering process is considerably lower to either dry rendering or wet rendering.

3.7.2 Unique Features of the Technology

The most striking feature of this package lies in the fact that

1. It is zero waste.
2. Economically viable for small throughputs.
3. Equipment needed can be fabricated locally; thereby no need for imports.
4. Easy maintenance of equipment.
5. Simplest possible in operation.
6. Low investment.
7. Environmentally clean.
8. Eco-friendly.
9. Use of local raw materials.
10. Production of high quality proteins locally.
11. Production of hides and skins without flaying defects.
12. Creation of employment even for the poorest of the poor.
13. Use of unskilled/semi-skilled manpower with just a week's training.
14. Doing away highly skilled manpower.
15. Low energy requirements based on locally available fuels.
16. Ability to work even if electricity is not available.
17. Ability to process offal from different sources like fishery, poultry, slaughterhouses, butcher shops, fallen stock etc.
18. Possibility of scaling up for large throughputs.
19. Establishing industries on raw materials available from rendering units.
20. And many more.

It therefore provides an excellent opportunity for tapping and utilization of vast resources of agriculture residues and animal based offal, which are hitherto wasted.

3.7.3 New Technology Package

While at the Central Leather Research Institute, the author had drawn up a plan of work to develop economically viable, environmentally clean and eco-friendly technology for the fallen carcass utilization as to be suitable under Indian rural conditions. Most aspects of planned work were investigated at the laboratory and thereafter, a field center was established at Bakshi ka Talab (BKT) near Lucknow, in the state of Uttar Pradesh in Northern India. The project was funded by the Department of Biotechnology, Government of India.

While this is so, some aspects of the work reported here, more particularly possible methodology for scaling up the technology, interpretation of the data on semi-moist basis and water activity were carried out by the author after his retirement.

The work plan followed at BKT is shown in the form of chart in Fig. 3.3.

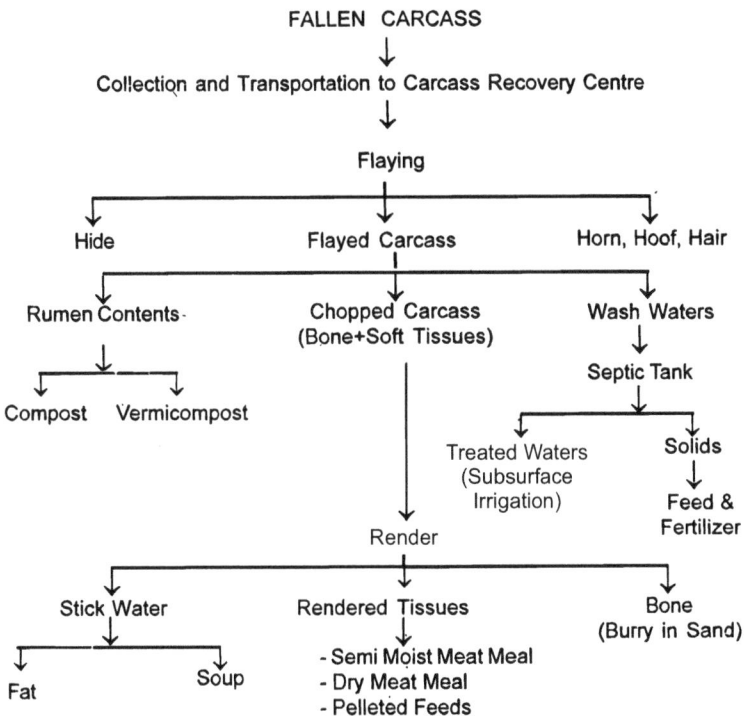

FALLEN CARCASS
↓
Collection and Transportation to Carcass Recovery Centre
↓
Flaying

| Hide | Flayed Carcass | Horn, Hoof, Hair |

Rumen Contents -　　Chopped Carcass　　Wash Waters
(Bone+Soft Tissues)

Compost　Vermicompost

Septic Tank

Treated Waters　　Solids
(Subsurface
Irrigation)　　Feed &
Fertilizer
Render

Stick Water　　Rendered Tissues　　Bone
(Burry in Sand)

Fat　　Soup　　- Semi Moist Meat Meal
- Dry Meat Meal
- Pelleted Feeds

Fig. 3.3 : Unit Operations for Fallen Carcass Utilization

As is seen from Fig. 3.3, there are six important aspects of successful utilization of fallen stock, which being

1. Collection of the carcass
2. Flaying of the carcass
3. Rendering of the flayed carcass for the production of meat meal/tallow etc.
4. Treatment and disposal of rumen contents
5. Treatment and disposal of effluents of all descriptions
6. Conservation of bone.

The manner in which work is carried out at BKT is discussed.

3.7.3.1 *Collection*

The collection of carcasses has conventionally been done by various means such as bullock cart, horse cart, mechanised vehicles. However an ideal transport-vehicle has to meet certain minimal requirements ; it has to be cheap, have low running cost and easy for loading and unloading of the carcass, on to and from it; more details are available in the literature.[8]

3.7.3.2 *Flaying*

After the animal is received at the flaying yard, it is hoisted on four legs (Figure 3.4) and then placed on the skinning bed (Figure 3.5) where it is ripped open and partly flayed.

Subsequently, the carcass is progressively hoisted upwards on its hind legs and flayed as it is raised. Manually operated chain pulley mounted over the over-head rail is used for hoisting and flaying of the carcass (Figure 3.6).

Proper ripping and flaying of the carcass partly on the flaying bed and partly on overhead rail helps in getting well flayed hides without flay cuts and in desirable square pattern. The flayed hide, the tail tuft and the earlobes (cattle), which had earlier been removed at the time of flaying, are suitably treated and preserved as raw materials. The details of collection of tail tuft as well as ear lobes are available in the literature.[3]

While the flayed carcass is still on rails, the entrails are removed into a wheelbarrow. The flayed carcass is now landed on to a wooden platform laid on the floor, where the same is chopped into smaller pieces with the help of an axe.

Fig. 3.4 : Carcass hoisted on four legs

Fig. 3.5 : Carcass being placed on skinning bed

Fig. 3.6 : Hoisted Carcass being flayed

3.7.3.3 Entrails

The stomach is cut off from other part of the entrails and the two separated. Thereafter, the stomach is transported to the compost pit, where it is cut open and rumen contents emptied in the pit. This is important because the flies prefer the smell of rumen contents over meat and so, flies if any, can be kept away from the flaying yard. The empty stomach is washed with water and transported back for rendering. It is desirable that stomach is chopped into small pieces before the same is rendered.

3.7.3.4 Rendering

The chopped carcasses as well as entrails are rendered in the semi-moist rendering vessel, the details of which are available in Annexure A.

The tissues are not only properly cooked and sterilized during rendering but also the meat and adhering sinews get detached from the bones; tallow is also separated or rather say extracted from the tissues during rendering. The horns and hoofs, which have been loosened from the pith and lamina respectively during the course of rendering are collected and dried[3] for storage.

The meat meal produced is of high quality; the protein content being around 65 per cent and fat being about 5 to 7 per cent on moisture free basis.

3.7.3.5 Mincing

After the raw material has been rendered, soft tissues are separated from the bones and minced with the help of meat mincer. After rendering, the soft tissues become so loose that their separation from the bones is possible with minimal efforts. Mincing helps in loosening the structure and reduction of size of the tissues, resulting in considerable increase in its surface area. More details on the role of mincing are available in Annexure-B.

3.7.3.6 Dry Meat Meal

Dry meal is produced when mince is dried. Many methods are available for drying such as hot air oven, rotary drier etc.[4] but drying in an open pan has been found quite satisfactory particularly when the throughputs are small (Figure 3.7).

Fig. 3.7 : Drying of the Mince in a Pan into Meat Meal

3.7.3.7 Semi-moist Meat Meal

Instead of drying the mince into dry meat meal, the same can be preserved wet as such into "semi-moist meat meal", which finds use in animal feeds in place of dry meat meal. The author has developed methods for the preparation of semi-moist meat meal by preserving the mince with the help of locally available cheap and edible materials like common salt and molasses. More information on the preparation of semi-moist meat meal and its water activity is available in Annexure–C.

3.7.3.8 Use of Semi-moist Meat in Pig Feed

In experiments conducts by the author, semi-moist meat meal in combination with other ingredients has been found to be an excellent additive in pig feeds; it had better palatability and gave higher conversion ratio in comparison to commercially available feeds. More details about pig feeding experiments are available in Annexure-D.

3.7.3.9 Pelleted Feeds

In yet another approach, rendered minced meat, in combination with other ingredients, is directly used for making pelleted feeds for fish and poultry etc.

The meat mincer itself with appropriate modifications is used for making pelleted feeds. Hence there is no extra investment on this account.

The mince and other ingredients like rice bran, wheat bran etc. are mixed in the form of dough; the soup (instead of water) generated in the process of rendering can be used for making the dough. In this manner, the soluble proteins present in the soup are put to good use. The dough is extruded (Figure 3.8) through the mincer in the form of pellets (Figure 3.9), which are dried or used as such in semi-moist condition itself.

3.7.3.10 Stick Water

Stick water is composed of 3 components, being (1) fat (2) water and (3) water soluble proteins; all the three released from the raw material during rendering. Water taken between "C" and "D" for initiation of the rendering operation also becomes part of stick water.

Fig. 3.8 : Dough being extruded in the form of pellets

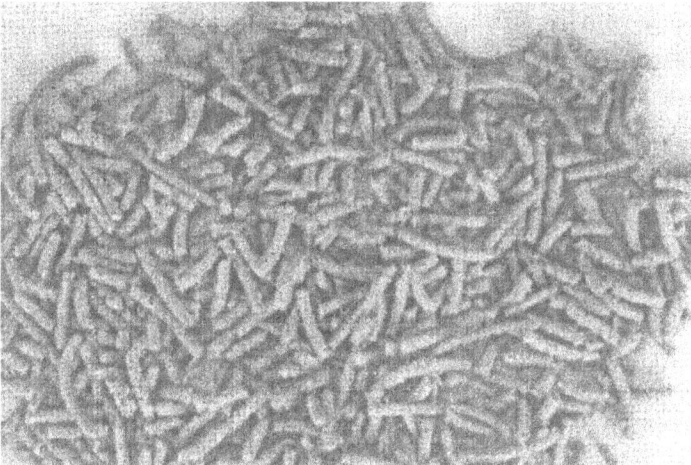

Fig. 3.9: Dry pelleted feed

The stick water is collected through valves "E" and "F" during and after rendering operation as explained in Annexure-A. The quality of fat in the stick water is of high grade, comparable to any best obtained in any other methods.

In fact, in the semi-moist rendering, the fat is progressively washed out or extracted out of the tissues as opposed to the wet rendering or dry rendering processes where it stays with the tissues throughout the rendering period; as a result, fat soluble colouring matter with the tissues dissolves in the fat. Consequently colour of the fat from the semi-moist rendering process tends to be whiter as compared to the other two processes.

The fat is separated from the stick water and refined following simple procedures.[2, 3] After fat has been removed, the remaining part consisting of water and soluble proteins can be rightly called "Soup" and as a matter of fact, used accordingly for feeding the pigs/piglets; it is also used for mixing ingredients for pellets into dough and pig feed ingredients into porridge like consistency. The soup contains about 4 to 6 per cent proteins.

3.7.3.11 Bone

The bone has been freed of fat during the process of rendering and separated from meat and sinews as discussed under 3.7.3.4 and 3.7.3.5, is still wet and needs to be suitably dried; this is done by burying the wet bones in a heap of dry sand. The sand absorbs moisture in few days and thereafter, these can be taken out of the sand and stored for sale. The sand can be used repeatedly for number of times before it is changed with new sand. This is simple, cheap and yet very effective method of drying rendered bones.

3.7.3.12 Compost and Horticulture

The rumen contents and so also any organic matter like weeds etc. growing at the centre are composted in compost bunkers (Figure 3.10).

The potential of compost has perhaps not been realized. According to estimates, if one cattle carcass is received every day and the rumen contents are composted, one will end up with sufficient compost to manure around 800 rose plants and 4,000 nursery plants (rose). At BKT, the manure produced is used at the center itself for maintaining a rose garden and a nursery. This has not only found proper disposal channel for rumen contents, but also added to the beauty of the centre and created some extra income. It is estimated that on the basis of rumen contents of just one cattle carcass a day, an average income of Rs. 100 (US $2) will be generated everyday from the sale of flowers and nursery plants. As such, an approximate income of around US $6 a day could be expected from the rumen contents of 3 to 4 carcasses a day. This will, no doubt, need extra land at the center for raising the garden.

Fig. 3.10 : Bunkers for Composting at BKT

Rose nursery and garden is just one example; depending upon local demand, compost can be alternatively used for raising other flowers or useful plants or sold as such.

Rumen contents can as well be converted into vermicompost; sufficient information is available on this subject. This will yield better income in comparison to simple compost.

Composting as well as vermicomposting has been comprehensively discussed in Chapter 10.

3.7.3.13 Treatment and Disposal of Effluents

The effluents from the flaying and rendering yards, urine, faeces and wash waters from the pig yard and so also night soil and urine from the toilets are sent to a specially designed septic tank with three compartments. The residence time of the effluents in the tank is a minimum of 20 days. In other words, the tank is so designed that effluents entering into it on a particular day, do not come out before 20 days. So treated outlet water is run into an underground channel filled with stone pebbles, which help the water to run to the end of the channel as designed. Trees like coconut, mango, lemon and many others can be planted along the banks of the channel, which get irrigated through seepage of water from the banks of the channel as seen in Figure 3.11, this system of irrigation is called as sub-surface irrigation. Root crops like carrot, radish and turnip should not be grown on these water.

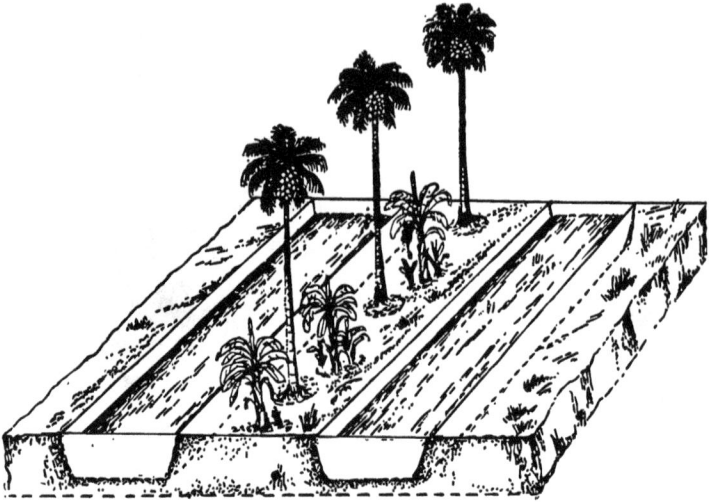

Fig. 3.11 : Sub-surface irrigation

Most of the organic matter in the effluents gets stabilized when retained in the septic tank and the bacterial flora is reduced considerably; as a matter of fact, the outlet water looks as clean as drinking water. However, so treated effluents are still rich in nitrogen, minerals and possibly many un-identified factors. As such, irrigation apart, the plants also receive some nutrients. As in the case of rose garden and nursery, this way of disposal of effluent is also expected to generate some income.

In addition, it has been seen (at BKT) that plants irrigated with the treated waters are very healthy, give high yield of crop and free of pest infestation. Consequently, it has been suspected that there may be certain un-identified factors in the treated waters, which help in keeping the pests away from plants.

Both the maintenance of a rose garden and nursery on the compost and various kinds of trees on treated waters has changed at BKT center into a garden with beauty and fragrance around. It is reported that plants absorb bad odors from atmosphere; it is thereby likely that bad odor, if any, emitted during dressing of the carcass is cleansed up by the plants around. The center is so clean that even house flies are rarely to be seen. This has added respectability to the place.

3.7.3.14 *Biogas Generation*

Biogas (methane) production has not been initiated at BKT. However, the author collected data on the quantities of faeces and urine generated by the pigs and potential of methane generation from this resource as well other effluents. In case, all the meat meal produced was to be used for raising pigs at the center itself, it is amazing to learn that an estimated 75 per cent of fuel needs (for rendering) can be met from this resource itself. In addition, the slurry produced during the process of biogas generation has been extensively used for irrigation and feed and fertilizer. The profile of the slurry has been studied and found safe for use in irrigation, feed or fertilizer as reported under 9.8.2, 9.8.3, 9.8.4 and 9.9.

Detailed information on the production of biogas is given in Chapter 9.

SEMI-MOIST RENDERING VESSEL: DESIGN AND OPERATION

The semi-moist rendering equipment used for rendering of the tissues was designed and reported earlier by the author[6]. The equipment has since undergone modification and is reported here. Photograph of the equipment is shown in Figure (Annex A-1).

The function of the equipment can better be understood with the help of its sectional view, which is shown in Figure (Annex A-2).

The equipment is a cylindrical metallic drum with bottom (C) and a circular removable plate (D), which rests on 3 angular supports as seen in the figure (Annex A-2). The outer surface of the drum has been heavily lagged with a suitable insulating material, to prevent loss of heat. The insulating material has been covered with a thin sheet of aluminum/stainless steel to avoid soakage of the same. The top end of the drum has been provided with a flange with a groove in which rests a circular rubber washer. The perforated plate "D" has been fitted with a perforated vertical pipe "B" as seen in the figure. A water level gauge "A" and draw off valves "E" and "F" have also been provided.

The vessel rests on a movable stand as shown in the figure; the same can be tilted sideways on the stand. This helps in easy loading and unloading of the tissues into and from the vessel.

The internal diamensions of the vessel fabricated and used at BKT as shown in Figure Annex A-1 were diameter 0.8 metres and height being 1.20 metres. This is equivalent to 600 litres of internal volume which accommodates roughly around ½ tonne of chopped

**Figure Annex A-1: Photograph of Semi-moist
Rendering Vessel**

raw material and found to be very convenient size for handling. The body wall thickness of the vessel can be 6 to 8 mm and that of base and dish end (lid) being 8 to 10 mm of SA516 grade 70 boiler quality steel plate.

The tilting of the vessel side ways, can be done either manually or mechanically for the size used at BKT (600 litres); manual tilting posed no problem. Mechanical tilting can be done by mounting a suitable worm gear at the tilting shaft and the right side stand of the vessel as can be seen in Figure Annex A-1.

After the rendering cycle is over, the vessel is to be opened and lid lifted. This can be done with the help of a manually operated chain pulley mounted overhead on a rail shown in Figure Annex A-1. Manual lifting of the lid is neither convenient nor advisable, being very heavy; manual lifting can lead to accident.

Small quantity of water (about 8 cm height) is filled in the space between C and D and the material to be rendered is accommodated in the space above plate "D". Soft tissues, carcass or parts thereof, have to be chopped into pieces not measuring more than 25 cm × 25 cm and loaded into the drum; the same is covered with a heavy lid fitted with a pressure gauge, a safety valve, a manually operated steam release valve and a thermometer. Steam could be released through the manually operated release valve if, for any reason, the pressure increases beyond the one set for the safety valve. The lid sits tightly over the rubber washer, which is secured by clamps fitted to the flange; C clamps are the most convenient for working. They can be opened or closed very fast as well as tightened easily.

The drum is fired from beneath as shown in the Figure (Annex A-2). The steam formed is forced through the perforated plate "D" and perforated pipe "B" and on to the material to be rendered. To begin with, the release valve is kept open to let the air and vapor escape; after all the air has been flushed out, the valve is closed.

The perforated plate "D" and pipe "B" facilitate the circulation of steam and water to all parts of the equipment. The boiling water rises through pipe "B" and strikes against hood "S", which redirects the water to fall on to the top of the material to be rendered; the water travels back to the bottom of the vessel and gets repeatedly circulated, as though it is a boiling water pump as could be seen in Figure (Annex A-3). Consequently, the material achieves the uniformity of temperature all through within a very short time.

As the temperature rises, meat coagulates at around 55°C and loses almost $2/3^{rd}$ of the water originally held by it; as a result, more water gets collected at the bottom, which in turn, starts soaking the tissues resting on the plate "D". For good results, the soaking of the tissues in water should be minimized. This is achieved by drawing extra water and fat through the draw off valve "F"; water level gauge helps in adjusting the level of the water desired to be maintained in the empty space between "C" and "D". Should the

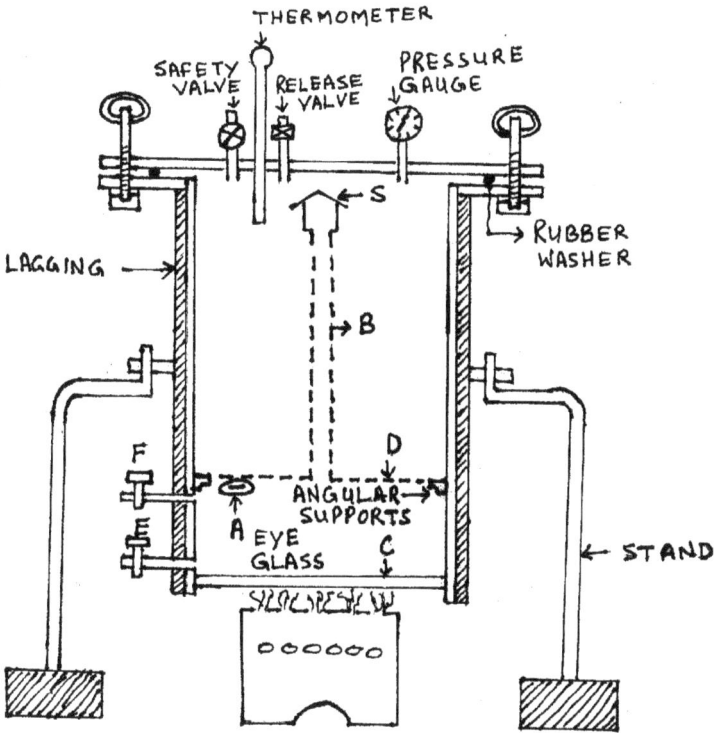

**Figure Annex A-2: Semi-moist Rendering
Vessel–Sectional View**

water be allowed to remain in contact with the tissues, there is abnormal softening and higher solubility of the same, which leads to lower quality and yield of the meat meal.

The rendering is carried out at temperatures varying between 115°C–145°C. After the desired pressure has reached inside the vessel, it is maintained at that upto half an hour and the fire is cut down thereafter, to the minimal; further cooking takes place with the heat locked inside (which has been made possible because of heavy insulation of the vessel as explained earlier) and also from the residual heat being supplied from the brazier. As a result, the fall of the temperature is very slow and gradual, which ensures

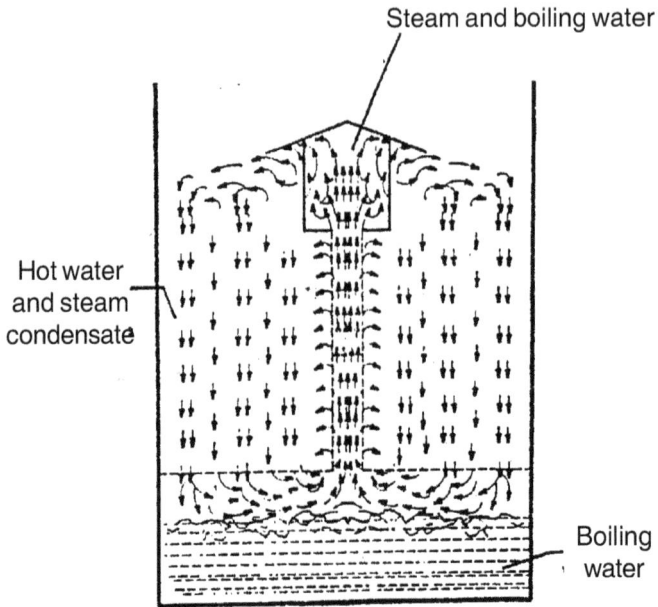

Figure Annex A-3: Semi-moist Rendering
Vessel–Water and Steam in Circulation

perfect sterility of the rendered material. If properly fired, the desired temperature of the material inside the vessel is achieved within 45–60 minutes.

Salmonella is the most problematic organism which the rendering industry faces. Fortunately, this gets destroyed if the material is exposed to 55°C for one hour or 60°C for 15–20 minutes.[13] Even so recontamination in subsequent operations like packing, transport etc. has to be avoided. This apart, the time and temperatures used for rendering inactivates all conventional pathogens and even reduces infectivity of the BSE agent by one to three logs (10– to 10,000 fold).[14]

Depending upon the weather, one cooking cycle takes about 8-10 hours. However, it has been found more practical to leave the contents overnight in the rendering vessel. The operation of loading and firing the vessel is normally started in the afternoon; the contents being left in the vessel overnight. The vessel is opened the

**Figure Annex A-4: Photograph of a Brazier Suitable
for Coal as Fuel**

next morning, stick water drawn out through valve E, the contents
taken out of the drum by tilting the same sidewise, meat separated
from bones (if applicable) and the same minced in a meat mincer.
As already stated elsewhere, the water content of the rendered
tissues stays at around 55 per cent at this stage; fat associated with
the raw tissues also gets extracted and washed out of the tissues
through repeated circulation of hot water and steam condensate

from top of the tissues to the bottom. The fat gets collected at the bottom of the vessel and drawn out through the draw off valves "F" and "E".

Any locally available fuel like coal, wood, biogas or agricultural residues could be used for firing the rendering vessel. Suitable designs of braziers are available for making use of different kinds of fuels; photograph of one suitable for coal as fuel, is shown in Figure (Annex A-4). In case electric power is available, the same can be used in place of the fuels mentioned; in such a case, the heating source will have to be incorporated in the design of the semi-moist rendering vessel.

MINCING

Rendering makes the tissues soft enough as to be minced with great ease; even the fibrous tissues like tendon and sinews could be minced effortlessly. A simple household 1H.P. meat mincer is able to mince around 50 kg of rendered tissues per hour. The mince (size being 5-6 mm) is dry to touch with open structure and does not get glued into lumps. This is for the fact that water was not allowed to remain in contact with the tissues.

It is important to mention that in places where power supply is either erratic or not available, the mincer can be operated manually with ease by directly attaching a flywheel to the moving worm shaft of the same. This makes the technology independent of power supply, which is very poor or not available in many developing countries. The photograph of a meat mincer in operation can be seen in Figure (Annex B-1).

Mincing is an important unit operation and plays vital role in the production of either dry or semi-moist meat meal as well as pelleted feed.

For converting the mince into dry meat meal, the same has to be dried. For small quantities, drying in an open pan has been found very satisfactory as has earlier been discussed under 3.7.3.7.

Meat is very bad conductor of heat—bigger the pieces longer it takes to dry. Mincing of the rendered stock not only reduces the size but also tremendously increases its surface area, which helps in faster drying into meat meal. While drying, the mince should be raked every now and then; raking is equally important for faster drying of the mince.

Figure Annex B-1: Meat Mincer in Operation

Mincing is also desirable for the preparation of either semi-moist meat meal or pelleted feeds as other ingredients of the feed can be easily and thoroughly mixed with the mince.

SEMI-MOIST MEAT MEAL AND ITS WATER ACTIVITY

Scott, in 1950's, defined water activity (represented by the symbol a_w) as the ratio of the water vapour pressure (P) over a food in a closed container to that over pure water (P_o) *i.e.*, saturation vapour pressure, at the same temperature

$$So, a_w = P/P_o$$

Thus, multiplication of the water activity by 100 gives the relative humidity (R.H.) of the atmosphere in equilibrium with the food.

$$R.H. (\%) = 100 \times a_w$$

When a solute like salt or sugar is added to water, the same (solute) binds with it (water) by increased molecular bonding. The bound part of water does not exert vapor pressure and thereby water activity of the solution is reduced. Bound water is also not available for bacterial activity. Since yeasts, molds or bacteria require a certain amount of available water to support their growth; non-availability (of sufficient water) will hinder their growth. Most bacteria do not grow at water activity below 0.910. Temperature, pH and several other factors can also influence whether an organism will grow in food and rate at which it will grow but water activity may be the most important factor.

The mince, as explained under Annexure-B, when preserved by controlling its water activity, has been termed as "Semi-moist Meat Meal." The author had earlier developed and reported formulations to control water activity (of the mince), using simple, cheap and locally available raw materials like common salt and molasses.[6] Two revised formulations, B and C, have since been developed and reported here. To get further insight, the moisture contents of B and C have been varied; moisture and water activity of all the samples measured and tabulated in Table (Annex C-1) and also in the form of graph in Figure (Annex C-1). Moisture content and water activity of the mince (A) as such have also been shown in Table (Annex C-1).

Table Annex C-1: Moisture Content and Water Activity

Sample ID	Description	a_w & Temp. in °C	Moisture
A	Mince[1] without additives	0.991/23.8	54.4%
B	Mince (100) + Common Salt (12)	0.854/25.7	50.5%
B_1	Moisture of B partly evaporated	0.806/24.7	44.5%
B_2	Moisture of B_1 further evaporated	0.747/25.4	33.4%
B_3	Some water added to B	0.869/24.8	51.2%
B_4	Further water added to B_3	0.883/26.3	54.5%
C	Mince (100) + Common Salt (10) + Molasses[2] (15)	0.875/25.8	48.0%
C_1	Moisture of C partly evaporated	0.841/25.6	41.9%
C_2	Moisture of C_1 further evaporated	0.719/26.0	27.1%
C_3	Some water added to C	0.883/26.0	49.1%
C_4	Further water added to C_3	0.898/25.6	51.6%

[1] 5mm; [2] 75% solids.

The effect of water activity on bacteria has been studied very thoroughly by various workers. As also discussed earlier, most bacteria do not grow at water activities below 0.910.[15, 16, 17] It is

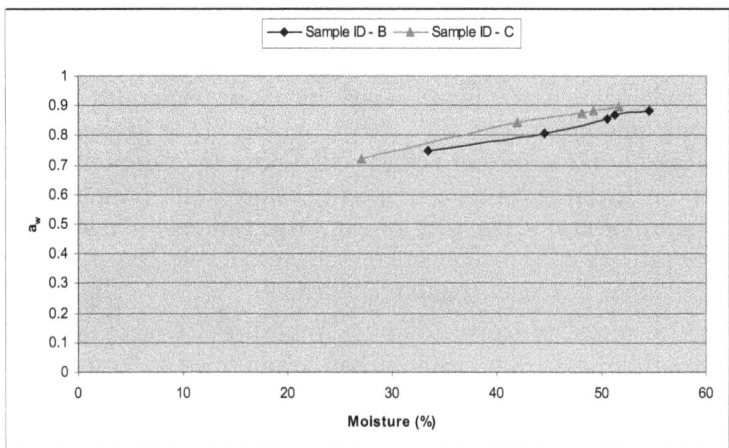

Figure Annex C-1: Moisture v/s Water Activity

seen from the table and the graph, the water activity of the semi-moist meat meals prepared (B through B_4 and C through C_4) is well below 0.910.

For preparation of the semi-moist meat meal, the solute/solutes are mixed thoroughly with the mince. The water activity of B or C being less than 0.910, these will preserve well as such, but it is likely that some contamination takes place during mincing and also mixing of the solute/solutes with the mince. To ensure complete sterility, it may be a good practice that the mix is subjected to a light heat treatment before it is packed airtight in a suitable container. In this form, the mix is sterile and has a long shelf life. The heat treatment is carried out in a pan with good raking of the mix while being heated; raking ensures thorough and uniform distribution of heat and sterility of the meal.

The chemicals used (salt and molasses) for controlling water activity are locally available in most of the developing countries at reasonable prices and are desirable components of the animal feeds. In other words, there is no real cost involved on chemicals, in the preservation of the mince; the salt or molasses could be suitably accounted for, while the feeds are formulated.

USE OF SEMI-MOIST MEAT MEAL IN PIG FEED

Rendered minced meat was prepared by rendering fallen carcasses in the semi-moist rendering vessel as reported under Annexure-A and preserved into semi-moist meat meal according to formulation B_1. The feed was prepared daily using semi-moist meal as one of the ingredients; the others being molasses, bone meal and from amongst the locally available agricultural produce and residues like rice bran, wheat bran, corn, sorghum etc. Soup from the stick water was used for mixing ingredients of the feed into porridge like consistency.

The age of the experimental piglets (Yorkshire) was 35 days at the start of the experiment. The feeding experiment was for a period of 12 weeks. The conversion of the feed by the experimental animals to weight gained at the end of the experiment was about 1:1 *i.e.* 1 kg of average weight gain for every 1.0 kg of feed consumed (on dry weight basis); this simple fact looks convincing enough of high profilability of the venture of pig raising.

The average weight of each piglet at the start of the experiment was 6.0 kg; the same being 40 kg at finish; this was more than 10 per cent higher than the control animals. Standard commercially available feed was used for control animals. The results are based on two batches of 10 animals each in each group (experimental and control).

The experiment was mainly conducted to explore the feasibility of feeding semi-moist meat meal in combination with other ingredients; no regard was given for balancing the protein and fiber content etc. of the feed in precise terms. As such, further work is needed to optimize the composition of the feed. The experiment establishes the feasibility of using semi-moist meat meal with very good results and without any problems; the feed was extremely palatable and did not create any indigestion or any other problems even for 35 days old piglets. The animals remained healthy and cheerful throughout the experimental period of 12 weeks and beyond thereafter.

The use of semi-moist meat meal for direct feeding brings down the cost of the feed as one saves on the inputs like labor, energy and infrastructure required for the purpose of drying of the mince into dry meat meal. The semi-moist meat meal can also be used in poultry and aquaculture feed formulations.

The idea of using semi-moist meat meal looks unconventional but it may be relevant to mention that ensiled fish meal in wet form is in regular use in cattle, poultry, mink and pig farm rations in Denmark for the past about 6 decades now. To be precise, the production of fish silage started in 1948.[18]

Feed apart, as mentioned under 3.7.3.10, soup or stick water obtained during the process of rendering can be used for feeding the pigs/piglets (as soup).

References

1. Ensminger, M.E., Oldfield, J.E. and Heinemann, W.W (1990): *Feeds and Nutrition*. Second edition. The Ensminger Publishing Co.

2. Mahendra Kumar (1989) : *Handbook of rural technology for processing of animal byproducts (FAO Agricultural Services Bulletin No.79)*. Food & Agriculture Organisation of the United Nations, Rome.

3. Mahendra Kumar (1987) : *Processing of animal byproducts in developing countries – a manual*. Commonwealth Science Council, Marlborough House, Pall Mall, London SW1Y 5HX.

4. Mann, I (1962) : *Processing and utilisation of animal byproducts (FAO Agricultural Development Paper No. 75)*. Food & Agriculture Organisation of the United Nations, Rome.

5. Scaria, K.J., Mahendra Kumar and Divakaran, S (1981): *Animal byproducts – their processing and utilisation*. NICLAI Publication – Central Leather Research Institute, Madras.

6. Mahendra Kumar & Carioca, J.O.B (2000): *Animal residues processing for feed ingredients, in "Recycling process for human food and animal feed from residues and resources"* edited by Carioca, J.O.B. and Arora, H.L., Banco do Nordeste and Universidale Federal do Ceara, UFC, Fortaleza - CE, Brazil.

7. Raghavan, K.V. and Thota, C.K (2001): *Rendering technologies – Scope of rendering processes, in "Technologies for Value realization of carcass byproducts in developing countries – a handbook"* edited by Rama Sami, T. Ranganayaki, M.D. and Rajagopal, N.R. Central Leather Research Institute, Chennai-600020 (India).

8. Mahendra Kumar (2007): *Animal byproducts utilization through semi-moist rendering.* Daya Publishing House, 1123/74, Deva Ram Park, Tri Nagar, Delhi-110035 (India).

9. Meeker, D.A (2007). *New book focusses on semi-moist rendering.* Render, October 2007, 2820 Birch Avenue, Caneino, CA 95709 (USA).

10. Burnham, F (1978) : *Rendering – the invisible industry.* Aero Publishers, Inc. 329 West Aviation Road, Fallbrook, CA 92028.

11. *Manual on Simple methods of meat preservation (1990) (FAO Animal Production and Health Paper No.79).* Food & Agriculture Organisation of the United Nations, Rome.

12. Mahendra Kumar, Rao, N.M., Srinivasan, T.S. Rose, C., Sehgal, P.K., Nigam, S.C., Kapahi, D.H. and Goel, D.K (1990) : *An approach for food production through fallen carcass utilisation and prevention of environmental pollution.* Pollution management in food industries (Page 71) in Proceedings of the symposium on impact of pollution in and from food industries, CFTRI, Mysore (India).

13. Franco, D.A (1993) : *Proceedings of the 54th Minnesota Nutrition Conference,* p. 21-35.

14. Taylor, D.M., Woodgate, S.L. and Atkinson, M.J (1995) : Inactiviation of Bovine Spongiform Encephalopathy Agent by rendering procedures. *Veterinary Record,* 137: 605–610.

15. Kuntz, Lynn A (1992) : *Keeping Microorganisms in Control.* Food Products Design, Aug. 92, pages 44–51.

16. Marsili, R (1993) : *Water activity: Why it is important and how to measure it.* Food Products Design, Dec. 93.

17. Grant A. Harris (1996) : *Food water relations.* Food Tech Europe, Dec. 1995/Jan. 1996.

18. Poul Hansen and Nile Alsted (1948) : Simple, low cost production of fish silage concentration. Ministry of Fisheries Technical University, Lyngby, Denmark.

CHAPTER 4

BONE

4.1 Composition

Bone is a hard tissue which gives shape and supports the structure of the body, provides attachment to the muscles and protects the vital organs of the thorax, abdomen as well as brain. Raw bone or the green bone, obtained fresh from slaughtered animal is composed of around 35–40 per cent water, 10–15 per cent red and yellow marrow and roughly 50 per cent solids as proteins and minerals.[1] The marrow consists of upto 96 per cent fat.[2]

Dry and fat free bones are composed of organic material and mineral matter roughly in the ratio of 1:2. Organic material mainly consists of the fibrous protein collagen; bone collagen is also called "ossein". The collagen is present in the form of matrix which is hard and calcified due to deposition of mineral matter (calcium salts). The mineral matter in itself consists of approximately 32.6 per cent calcium and 15.2 per cent phosphorous. Small amounts of sodium, potassium, magnesium and traces of zinc, copper, iron, cobalt, sulphur and manganese are also present.[2,3,4]

4.2 Yield

A number of factors like age, sex, breed, state of health, nutritional status etc. of the animal, affect the yield of bone. In beasts in top condition, it may be as low as 12 per cent and as high as 30 per cent in emaciated animals based on the live weight of the animal.

4.3 Availability of Bones in Developing Countries

Generally speaking, bones may be classified into the following three categories:

1. Green bones
2. Table bones
3. Desert bones

Green bones are the ones obtained from freshly killed animals and which contain high percentage of moisture and fat. It may be said that not much of green bones are available in the developing countries. Meat is usually sold along with the bones and the two are cooked together for edible purposes. At times, legs may be deboned to extract some long bones at the meat vendors shop but these again may be sold as soup stock. In some countries, skull and jaw bones are available at the slaughterhouse while in others, only jaw bones may be available. Still in others, even jaw bones are not available at the slaughterhouse.

The table bones are the ones which have already been cooked either along with the meat or as soup stock and are the left overs after meal from hotels, eating places and houses. There is no organised collection of this resource and the same is usually thrown as garbage. If collected, table bones will be available in big quantities in the developing countries.

Fallen carcasses are the main source of the so called "desert bones". These have been exposed to sun, rain and weather for long periods and this exposure frees them of fat and adhering meat etc. They have also dried over the period. As such, these are clean bones, light in weight, consisting of calcium, phosphorous, other minerals in small amounts and the collagen matrix – the ossein. At times, some sinews may also be adhering with such bones. Mortality of livestock is high in developing countries and so, large quantities of bones are available from this resource in these countries.

4.4 Uses of Bone

Bone is a special raw material in as much as it contains both organic and inorganic constituents in appreciable proportions. Consequently, varied types of products could be produced from bones. Many types of industries which could be established based on bone are as follows :

1. Handicrafts
2. Fat extraction

3. Bone crushing

4. Mineral (calcium + phosphorous) based products

5. Protein (ossein) based products

6. Bone char

Considerable information is available on the production of all these products and one is advised to refer to the literature for more details.[3,4,5] However, the following three products will be discussed in detail in this chapter.

1. Handicrafts

2. Bone meal

3. Bone ash

This is for the fact that handicrafts ensure the best possible return from among the bone products and many countries could benefit from this industry – in particular the ones which have a high tourist inflow like Kenya, Philippines, Thailand and India to mention only a few. Such products have a high potential to be sold as souvenirs. The other two products viz. bone meal and bone ash are simple to be produced and have immediate utility both in agriculture and live stock feeds.

4.5 Handicrafts

A variety of products, having both aesthetic appeal and utility are produced from bones, such as jewellery, inlay works, pen holder, flower vase, name plates, candle stand, ash tray and the like.

There cannot be an exhaustive list of such articles because possibilities are innumerable. It could be left to the imagination of the artist to carve out more and more products of different contours and dimensions; the criteria is an article of beauty preferably with some utility.

Both green as well as desert bones could be used for handicrafts. In either case, the bone has to be intact with proper sheen and undamaged surface. Green bones have to be treated suitably to free them of the fat and adhering tissues before use. A discussion on the structure of the bone, will help in the understanding of the treatment method for defatting etc. in its proper perspective.

Nowadays, plastics have replaced the use of bone to some extent in the manufacture of such handicrafts. However, enough market still exists for genuine goods made out of bone particularly from the tourists from affluent societies. There is a great love and appreciation for handicrafts in these societies as hand made goods are not made in their own countries. A variety of these types of handicrafts are made in countries like India and Philippines. Other countries could develop this industry based on their local skills. Handicrafts offer considerable employment opportunities.

4.5.1 Dense and Spongy Bones

Structurally, bones are of two types – dense and spongy. In the spongy bone, the marrow cavities are large and irregularly arranged. The bony substance consists of large, slender spicules called "trabeculae". The "epiphysis" (at the end of the shaft of a long bone) consists of spongy bone (Fig. 4.1). In the dense compact bone, such as the tibia, the marrow cavities are narrow and the bony substance is densely packed with cells and matrix (Fig. 4.1). Marrow contains as high as 96 per cent fat.

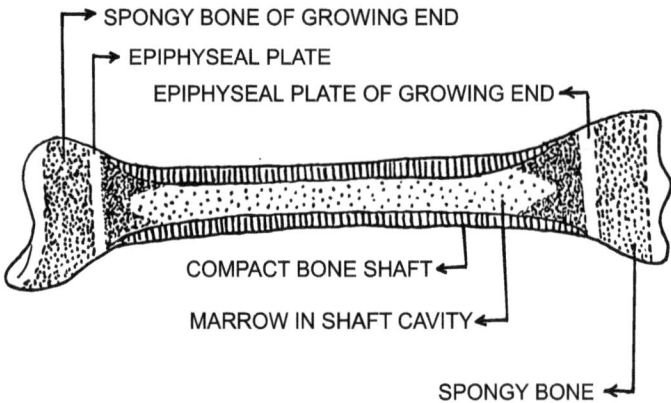

SPONGY BONE OF GROWING END
EPIPHYSEAL PLATE
EPIPHYSEAL PLATE OF GROWING END
COMPACT BONE SHAFT
MARROW IN SHAFT CAVITY
SPONGY BONE

Fig. 4.1 : Dense and spongy parts of bone.

From the point of view of handicrafts manufacture, it is the dense bone which are more valuable. Six pairs of leg bone viz. thigh bones (tibia), buttock bones (femur), flat (metatarsus) and round (metacarpus) shin bones, blades (radius and ulna) and cannon (humerus) bones fall under the category of dense bones. Spongy bones find lesser use in handicrafts; their use being almost limited to inlay works. Rib bones which are spongy, are used for a variety of inlay works.

4.5.2 Treatment

As mentioned earlier, the treatment is aimed to free the bone of fat and adhering matter. This is normally done by boiling the bone in hot water.

The marrow as could be seen from the structure under Fig. 4.1, is packed inside in a narrow cavity. If the bone is boiled with its structure intact, it is not possible to reach the marrow and so it's removal is not possible. Hence the bone has to be suitably prepared before the same is put to boiling treatment. The preparation depends upon the very object one plans to make. For example, for a pen holder one or two holes are drilled in a suitable place for holding the pen (Fig. 4.2). Similarly, the bone is suitably cut and laid for a inlayed table (Fig. 4.3). A power driven circular saw is the best equipment for cutting off the bone for this purpose. After the bone has either been cut or holes drilled, marrow could easily by exposed to the boiling water.

For boiling, the bones are loaded in a suitable pan along with cold water and the contents brought to boiling at a slow rate. Bone should never be placed directly in boiling water; this may cause cracks.

About 8–10 hours treatment may be needed for the fat to separate properly from the bone. Bones of older animals require more cooking time than that of younger ones. The time of cooking has to be properly adjusted. Undercooking leaves greasy bones while overcooking makes them brittle.[2,3,4]

After the cooking is over, the contents in the pan are allowed to cool (to around 60°C) and settle. The fat layer separates at the top. The same is skimmed off. Fat is used for a variety of purposes such as in livestock feeds, candle and soap making and in many other industrial applications.[6,7]

The treated bones are taken out from the pan, brushed to get rid of any adhering matter, given one or two washes with warm water and thereafter dried on a cemented platform in the open sun for about a week. So treated material is now free of any fat, adherents and odour and ready for use for handicrafts.

Desert bones may also be treated similarly before these are used for handicrafts but the time of treatment may be much less when compared to green bones.

4.5.3 *Preparation of Handicrafts*

The bone is suitably chosen, cut into desired shape of the product planned to be produced and then polished to get a fine smooth surface. Finally, the finished surface may be given a coat of a transparent lacquer and the coat dried. The coat protects the finished surface and gives shine and smoothness.

Simple tools like circular saw, hack-saw, drill, files and emery paper etc., are required for the production of bone handicrafts.

4.5.4 *Quality of Finish*

Quality of finish is greatly dependent upon the animal from which the bone has been derived. For example, camel bone takes smoother and finer finish when compared to cattle bone. For this reason, quality products are preferably made from camel bone. A number of bone based handicrafts are shown in Figs. 4.2 to 4.7.

4.6 Bone meal

Bone meal, as the name suggests, is a ground powder of definite fineness, made out of bone; the bone may be raw or subjected to treatment before grounding. The Bureau of Indian Standards has drawn specification for three types of bone meal. The same will be described. Analytical procedures for evaluation of the products have also been laid down in the respective standards.

4.6.1 *Bone Meal, Raw*

Bone meal (raw), is a product normally obtained in the process of bone crushing – as much as upto 25 per cent of the total bone crushed may be obtained as bone meal (raw).[3,4] This product is normally used as fertilizer. Care should be taken that bones derived for crushing have not originated from animals suffering from Zoonotic diseases – anthrax in particular. The danger of communication of diseases to humans while handling such bone could be well imagined; this aspect has been discussed at some length in Chapter 11.

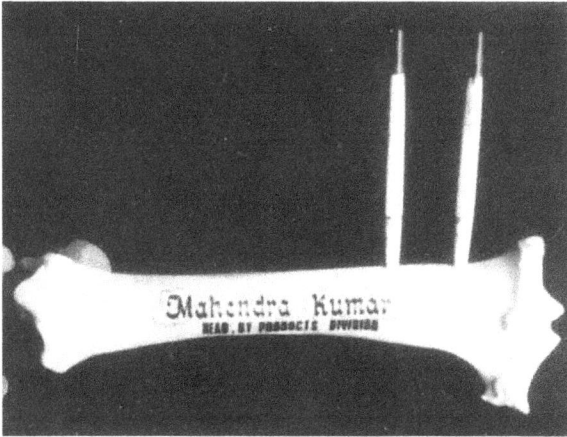

Fig. 4.2 : A pen holder with name printed (Buffalo bone).

Fig. 4.3 : An inlayed table

Fig. 4.4 : Inlayed figure of an elephant.

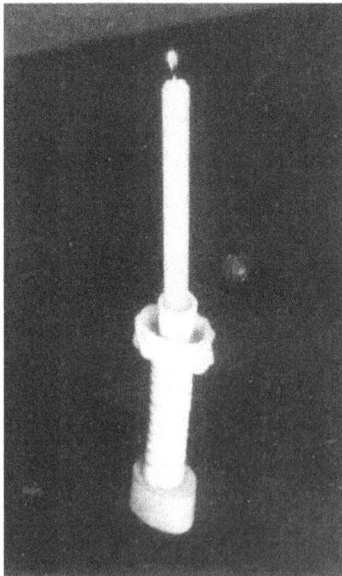

Fig. 4.5 : A candle stand (Camel bone).

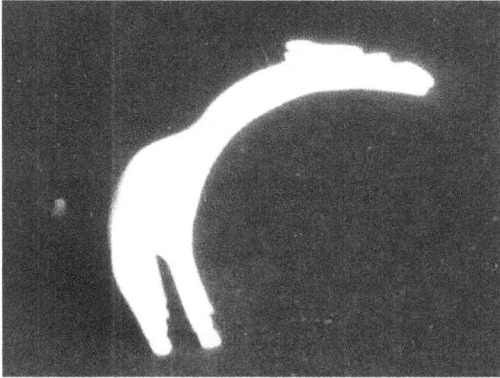

Fig. 4.6 : A figure of a giraffe (Camel bone).

Fig. 4.7 : Jewellery articles (Camel bone).

The following standards have been laid down for Bone Meal, raw.[8]

i	Moisture, per cent by weight	Max	8.0
ii	Acid insoluble matter, per cent by weight	Max	12.0
iii	Total phosphates (as P_2O_5), per cent by weight	Min	20.0
iv	Available phosphates (as P_2O_5), soluble in 2 per cent citric acid, per cent by weight	Min	8.0
v	Nitrogen content of water insoluble portion, per cent by weight	Min	3.0

4.6.2 Bone Meal as Livestock Feed Supplement

This is a product which has been prepared by defatting and sterilizing of undecomposed bones by steam under pressure. The bones are loaded in a bone digester (Fig. 4.8) and subjected to a steam pressure of 2.8 kg/cm^2 (40 psi) for a period of about 2½ hours. Higher pressures have also been used; in that case, time of exposure under steam gets correspondingly reduced. The bone gets sterilized, becomes soft and some protein and fat gets extracted during the process of steaming; the extract gets collected at the bottom of the bone digester, which is later drawn out through the valve provided for the purpose. After steaming, the pressure in the digester is allowed to drop to the atmospheric pressure, bone taken out, dried in the sun and milled to desired fineness.[2,3,4]

The specifications laid down for this product are as follows.[9]

i	Moisture, per cent by weight,	Max	7.0
ii	Calcium, per cent by weight,	Min	31.0
iii	Phosphorous, per cent by weight,	Min	14.0
iv	Crude fat, per cent by weight,	Max	1.0
v	Total ash, per cent by weight,	Min	85.0
vi	Acid insoluble ash, per cent by weight,	Max	.06
vii	Spores of *Basillus anthracis* and *Clostridium*, sp.	Nil	

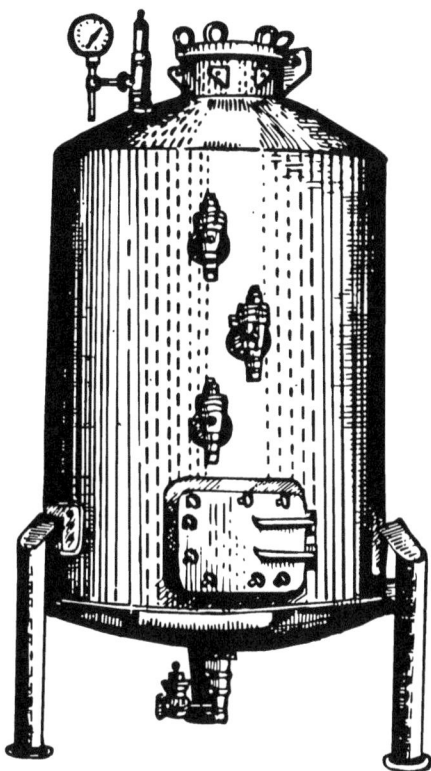

Fig. 4.8 : A bone digester

4.6.3 Bone Meal-steamed

Bone meal (steamed), also called as Steamed Digested Bone Meal is prepared by prolonged steaming of undecomposed bones immersed in water in a bone digester under pressure; the fat, glue and nitrogenous matter gets extracted in this process. As a result, the product predominantly consists of mineral matter with very little of protein left unextracted. In the other product discussed under 4.6.2, which also has been prepared by steaming, only a small percentage of protein is extracted, resulting in a product of comparatively higher percentage of protein content and lower percentage of mineral content.

The standard prescribed for Bone Meal (steamed) is given below.[10]

i   Free moisture, per cent by weight   Max   7.0

ii   Total phosphates (as P_2O_5), per cent
by weight (dry basis)   Min   22.0

iii   Available phosphates (as P_2O_5),
soluble in 2 per cent citric acid
solution, per cent by weight (dry basis) Min   16.0

Glue manufacture from bone is also based on the principle of deproteinisation of the bone. The deproteinised bone obtained in the process of glue manufacture meets with all the requirements laid down for bone meal (steamed), the process details are available in the literature.[3,4]

4.7 Bone Ash

Bone ash, also called as "calcined bone" is prepared by burning the bone where total combustion of the same is ensured.[2,3,4]

Depending upon the quantity of the raw material available, different types of calcining systems are followed. For very large quantities, rotary kilns like the ones used in cement industry are used. Conventional kilns used for firing lime stone and pottery are also used. Yet in another method, bones could be converted into ash by burning them by a very simple method. The method could be followed even in villages and far flung areas without any difficulty. To begin with, the method could be adopted in such of those places where bones are not alternatively utilized. The method will be explained in detail. Bone ash finds use both in agriculture and livestock feeds.

A base grill measuring about 125 cm × 125 cm is made out of iron bars; the bars have a spacing of approximately 10 cms. The base grill is provided with grilled four side walls, each measuring about 20 cm. in height. The grill is made to rest on two brick walls of about 40 cm height from the ground (Fig 4.9).

The bones are piled on the grill to a maximum height of 20 cm. Long and dense bones are placed at the bottom while the small, flat and spongy ones on the top of the pile (Fig. 4.9). Bones of all description, i.e. desert bones, table bones and green bones could be utilized for making bone ash. Green and table bones should be dried for 2-3 days before they are used for calcining. Care should also be taken that bones are free of mud, grit, sand etc. before calcining. This ensures purity of the calcined product.

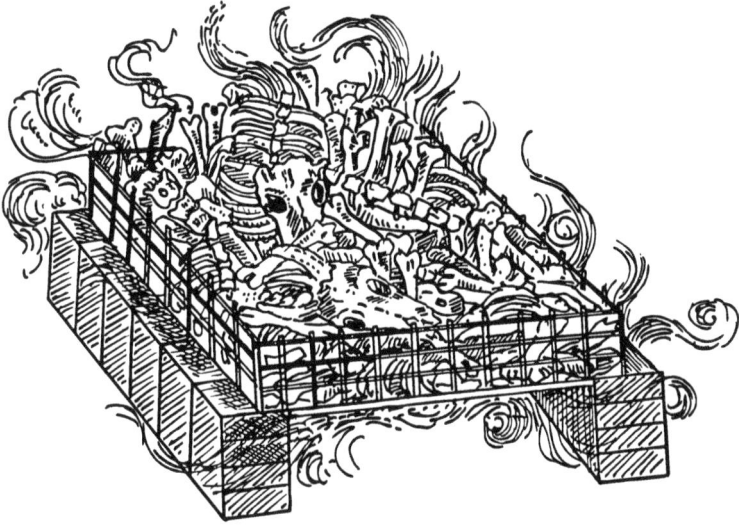

Fig. 4.9 : Grill resting on brick walls, loaded with bones and fired

A small quantity of kerosene is sprinkled on the pile and the same set to fire. Once the bones catch fire, they burn well on their own because of fat, ossein, sinews and meat tissues associated with them. The process of burning takes about 1-2 hours. If kerosene is not available, the bones are set to fire by burning fire wood underneath the grill.[2,3,4]

Properly calcined bones become white to pale yellow in colour but retain their original shape and contours. They, however, become so soft that it becomes very easy to powder them by pounding, or in a mill. The ones burnt partially are black and comparatively hard. These are returned back to the next batch for their complete burning.

Calcining could also be carried out underground instead of on the ground. In that case, the pile as explained, is arranged in a trench. Being underground, loss of heat is minimised. This helps to save on fuel costs.[2,3,4]

The calcined product gets completely devoid of organic matter and is sterile. It contains over 15 per cent phosphorous and approximately 31 per cent calcium. The yield of bone ash is around 65 per cent on dry weight of the bones, when the raw material used is dry and free from other contaminants like sand, grit and mud. Indian standards has laid down specifications for calcined bone which are as follows.[11]

i.	Moisture, per cent by mass	Max	0.50
ii.	Calcium, per cent by mass	Min	31.00
iii.	Phosphorous, per cent by mass	Min	16.00
iv.	Fluorine, per cent by mass	Max	0.05
v.	Acid insoluble ash, per cent by mass	Max	3.00

4.8 Uses

The most popular use of bone meal and calcined bone is in agriculture, livestock feeds and manufacture of mineral mix for livestock feeds. This apart, bone ash is also used for the manufacture of dicalcium phosphate and Bone China – a high class porcelain valued for its lightness and translucency, and as a burnishing agent.

In agriculture, the products are used for the supply of phosphorous to the plants. While this is so, it is most desirable to use the bone products in acidic soils, wherein the calcium helps in the correction of acidity of the soil. Bones of all descriptions and particularly the low grade materials like jaw bones can be used for products to be used in agriculture.

For the manufacture of mineral mix for use in livestock feed, both calcium and phosphorous present in bone products are equally important. Products like steamed digested bone meal and bone ash are preferred for use over bone meal in the manufacture of mineral mix. This is for the fact, that weight to weight, these products have more minerals as discussed under 4.6.3. Whichever is the product used, it is important that the same is sterile or at least absolutely free from pathogens. The fluorine content of the products meant for livestock feeds is to be kept as low as possible. Jaw bones have higher fluorine content than other bones. Hence, use of jaw bones in livestock feed products is avoided. Jaw bones are usually diverted for products used as fertilizers. [3,4]

Bone ash, meant to be used for Bone China, is to be made from good quality bones free from external contamination, and iron in particular. Iron interferes with the colour and translucency of the Bone China.

4.9 Other Bone Based Products

Many other products such as ossein, dicalcium phosphate, glue, gelatine and bone char are produced from bones. Detailed information is available on the production of these products in the literature.[3,4,5]

References

1. Gelatine (1984) : *An overview of the world market with special reference to the potential for developing countries.* International Trade Centre, UNCTAD/GATT, Geneva.

2. Mann, I (1962) : *Processing and utilization of animal byproducts (FAO Agricultural Development Paper No. 75).* Food and Agriculture Organisation of the United Nations, Rome.

3. Mahendra Kumar (1989) : *Handbook of rural technology for processing of animal byproducts (FAO Agricultural services Bulletin No.79).* Food and Agriculture Organisation of the United Nations, Rome.

4. Mahendra Kumar (1987) : *Processing of animal byproducts in developing countries – a manual.* Commonwealth Science Council, Marlborough House, Pall Mall, London SW1Y 5HX.

5. Sastry, T.P. and Rao N.M (2001): *Processing of bones, in "Technologies for value realization of carcass byproducts in developing countries – a handbook"* edited by Ramasami, T. Ranganayaki, M.D. and Rajagopal, N.R. Central Leather Research Institute, Chennai-600020 (India).

6. Burnham, F (1978) : *Rendering – the invisible industry.* Aero Publishers, Inc. 329 West Aviation Road, Fallbrook, CA 92028.

7. Filstrup, P (1976) : *Handbook for the meat byproducts industry.* Alfa-Laval Slaughterhouse Byproducts Department, Titan Separator A/S, Denmark.

8. IS : 853-1964 : *Specification for bone meal, raw (revised, reaffirmed 1984).*

9. IS : 1942-1968 : *Specification for bone meal, as livestock feed supplement.*

10. IS : 1014-1956 : *Specification for bone meal, steamed.*

11. IS : 7061-1973 : *Specification for calcined bone-meal as livestock feed supplement.*

CHAPTER 5

BLOOD

5.1 Introduction

Blood is an opaque fluid, red in colour when oxygenated and purplish when not oxygenated. When the animal is slaughtered, blood gushes out from the blood vessels. The recovery of raw blood after slaughter of the animal varies between 3 to 5 per cent on the live weight of pigs, cattle, sheep and goat.

The solids present in blood are almost comparable to that of meat; veal, for example, may contain 20 per cent solids, the corresponding figure for blood being around 18 per cent. About 90 per cent of blood solids are proteins composed of haemoglobin, albumin, globulin and fibrinogen; the other constituents of blood, in small fractions, are sodium, potassium, calcium, magnesium, iron, chloride, phosphoric acid, sugar, cholestrol, fat and lecithin.[1,2]

Blood finds many uses – in food, feed, fertilizers, pharmaceuticals and industry[3,4,5]; the discussions here will be restricted to its use in feeds under the conditions of developing countries in particular.

5.2 Collection of Blood

Collection of blood, after the animal has been slaughtered, is done by either of the following two methods

1. On rail

2. On floor

The former method is the desirable one; it ensures complete bleeding of the animal and collection is made possible in a more hygienic manner. Unfortunately, however, the collection is most often done on floor in most of the developing countries.

5.2.1 Collection on Rail

In this method, the animal has to be hoisted on rail; for doing so, the animal should be stunned. Stunning makes the animal senseless, which in turn, makes hoisting easy.

Stunning is done by various methods viz. electrical stunning method[6], captive pistol bolt method[7] and by piercing the brain with a sharp pointed knife (Fig. 5.1)

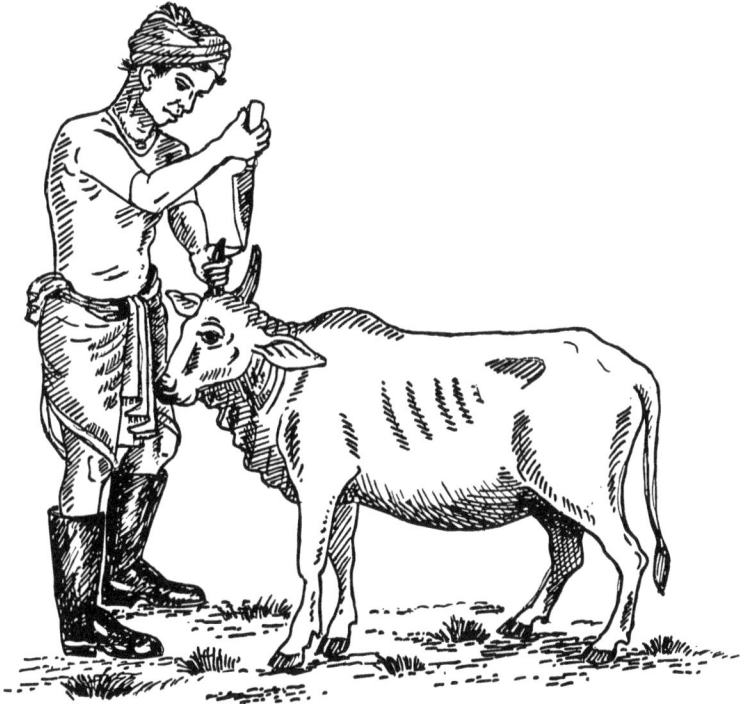

Fig. 5.1 : Stunning with a sharp knife. An experienced worker is able to reach the brain and puncture the same in one single powerful stroke.

Soon after the animal has been stunned, the same is hoisted where it is slaughtered and allowed to bleed. In organised abattoirs, blood is allowed to drain into an underground tank wherefrom the same could be pumped for its processing (Fig. 5.2)

Fig. 5.2 : Carcass hoisted on rail and bled.

Should the number of animals to be slaughtered be small, the same hoisting system could be followed and blood collected in an oil drum suitably modified for collection. The drum is cut from the side to accommodate the head of the beast; this helps in complete bleeding of the carcass into the drum (Fig. 5.3).

In this method, bleeding takes place against gravitational force and the carcass bleeds optimally and completely, ensuring highest possible collection.

Fig. 5.3 : Carcass being bled into an oil drum.

In places where rails are not possible to be installed, improvised hoist systems could be made using local materials like bamboo or wooden posts. A tripod structure made out of bamboo posts, on which a hoist has been secured, is shown in Fig. 5.4.

Both electrically and manually operated hoists are available. Electrical hoists are expensive and their installation cannot be justified when operations are small and where electrical failures are frequent. In fact, electrically operated hoists have not met with success in many developing countries. Manually operated hoists are cheap and independent of power supply. Their maintenance as well as operation is simple and could be operated by single individual for hoisting the carcass. As such, installation of manually operated hoists (Fig. 5.5) is recommended particularly when operations are small. A one tonne hoist is quite suitable for the job.

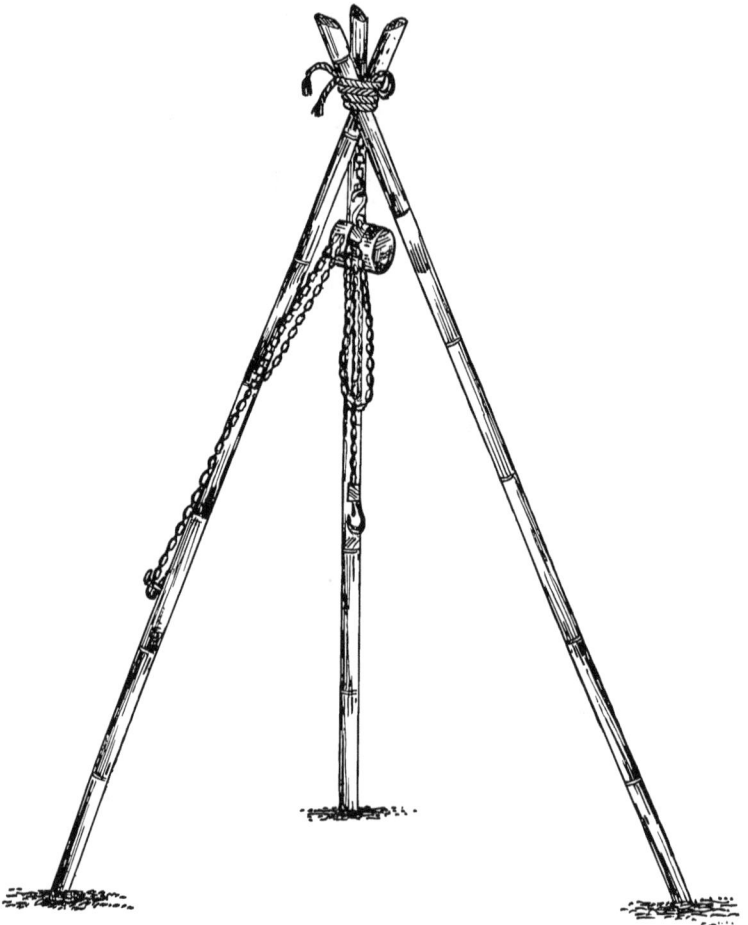

Fig. 5.4: An improvised hoist system.

Fig. 5.5 : A Manually operated hoist.

5.2.2 *Collection on Floor*

As discussed earlier, in most of the developing countries, the animals are slaughtered on the floor and rarely, if any, on rail.

The animal is made to fall on the floor by entangling either the pair of hind or forelegs with a rope and pulling the same side way. As the animal falls, its four legs are firmly tied with the rope. In the case of small animals like sheep and goat, the animal is not made to fall on the floor; the movements are controlled by holding the animal firmly before it is slaughtered.

For the collection of blood, a metallic pan is kept right below the neck of the animal whereafter the animal is slaughtered. As the blood gushes out, it gets collected into the pan (Fig. 5.6).

The quantity of blood which is possible to be collected is almost half compared to that collected on rail. This happens because of two reasons; firstly, the animal does not bleed properly while lying on floor and secondly, some blood always, spills outside the pan as the animal struggles after slaughter. In fact, to minimise the spillage, the head of the animal should be firmly held in position so that collection of the blood into the pan is ensured. The pan is a metallic oval shaped vessel, made out of galvanised tin sheet, measuring about 60 cm in length, 50 cm in breadth and 15 cm in depth. Two handles are provided on either side for easy handling of the same.

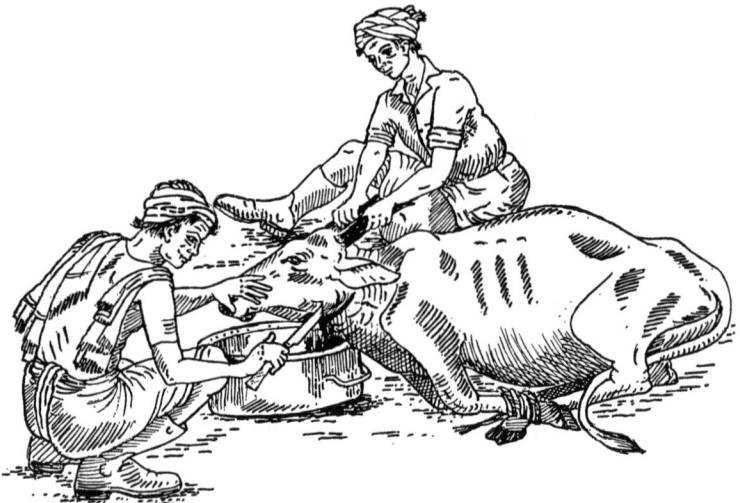

Fig. 5.6 : Blood collection on floor.

The method as explained, is very torturous and painful to the animal and should give way to humane method of killing. Accordingly, the animal should first be stunned and slaughtered thereafter. More information on killing and collection of blood on floor is available in the literature.[1,2]

Irrespective of the method to be followed, the following general precautions should be taken while collecting blood.

1. The collection should be done in as hygienic a manner as possible.

2. Blood should not be allowed to get mixed with water; this is likely to contaminate blood and will mean additional cost in its processing into blood meal.

3. Mixing of blood with any other foreign matter should also be avoided. The most common contaminant is the undigested food from the stomach through the vomiting while the carcass is bled. This could be prevented quite easily by grasping the oesophagus (gullet) firmly in one hand at the time of bleeding. Alternatively, a metal clamp is used to block the gullet in place of hand.[1,2,7]

4. The blood collected should originate from inspected animals and processed without loss of time after collection.

5.3 Processing of Blood

Processing of blood as animal stock feed, for small throughputs alone, will be discussed here. The following three products could be made :

1. Blood meal
2. Lime treated blood
3. Absorbed blood

5.3.1 Blood Meal

Production of blood meal by very simple method is possible. The process consists in cooking of the blood in a pan on direct fire (as cooking of meat Fig. 3.2) ; heating should be done with constant stirring until the contents are converted into black mass. Charring must be avoided; continuous stirring helps to achieve this. The contents should boil for about 20 minutes to complete coagulation and achieve sterilization of the cooked mass. In case steam is available, coagulation as well as sterilization could be achieved by injecting steam into the blood. In this method, charring can be avoided completely.

The cooked mass is to be dried into blood meal; the process normally involves in pressing out excess moisture from the mass and thereafter drying the cake into blood meal. Drying of the cake in the direct sun on a cemented platform (as in the case of rumen contents – Fig. 6.1) is possible; alternative methods are also available. For details one is advised to refer to the literature.[1,2,7,8]

5.3.2 Lime Treated Blood

Lime when mixed with blood, helps in its preservation. This is a very simple way of saving blood and can be followed without any problem even in the remote places.

The process consists in taking finely powdered unslaked lime or quick lime (1 per cent on the weight of blood) in the blood collecting pan and then collect the blood directly into the pan from the bleeding slaughtered animal. After the requisite quantity has been collected, the contents are stirred well until the mix sets into a black, rubber like mass. When quick lime is not available, 3 per cent powdered slaked lime can be used but the process of setting is slower than with quick lime.

The lime treated blood, in the ratio as explained, keeps well without spoilage for about 7-8 days.[1,2,7] If still higher percentage of lime is used, the mix can be kept without spoilage for upto 2 months[9]. This however, may not be found necessary as the treated blood as explained will normally dry within 4-5 day even in high humid climates.

Drying of the treated mass can be done in direct sun on a cemented platform, as in the case of rumen contents. Lime treated blood does not stick to the surface, nor does it attract flies. The product dries into crumbs which have to be powdered before use.

The product could be used in livestock feeds – pigs and poultry in particular. Addition of lime to blood does not in any way interfere with its nutrition value. Of course, lime should be accounted for, as the feed is formulated.

5.3.3 Absorbed Blood

Like treatment of blood with lime, this is yet another simple method of conservation of blood particularly when sun is available. The process is simple and amounts to mixing of the blood with a suitable absorbent and drying the mix in the sun. Any material which has absorbent property and is an accepted ingredient to be used in the livestock feeds, could be utilized for the purpose. Typical examples of some of the absorbents are, dry rumen contents (R.C.), wheat bran and rice bran.

Sun dried R.C. is an excellent absorbent in as much as it absorbs the highest quantity of blood and the absorbed mass dries in the shortest possible time.[1,2] This is illustrated in Table 5.1.

The criteria for deciding the quantity of blood to be mixed with the absorbent is dependent upon the handling property of the mix; the consistency of the mix should be satisfactory both for handling as well as spreading of the same on the cement platform. The consistency has got to be just right – neither too dry nor too wet to handle.

Table 5.1 : Absorbent power of dry R.C. and wheat bran and time required for drying of the mix

Item	Quantity of absorbent	Quantity of blood	Time required for drying of the mix*
Dry R.C.	1 Kg.	1.50 Kg.	4-5 hrs.
Wheat bran	1 Kg.	0.75 Kg.	10-12 hrs.

* The ingredients mixed thoroughly and the mix dried under identical conditions.

In actual practice, dry R.C. is spread over a polythene sheet and blood collected from the slaughtered animal is poured over the same and the two mixed thoroughly by hand; care should be taken that blood does not coagulate before it is mixed with the absorbent. This is possible; blood starts coagulating after about 3 minutes from the time the carcass starts bleeding. It takes less than 1 minute for the carcass to bleed completely.[1,2] Hence, clear 2 minutes are available for the blood to be mixed with dry R.C. before the same starts coagulating. While working in Africa, the author did not have any difficulty in handling blood from upto 10 cattle a day in this fashion. The mixing of the blood and absorbent should be thorough such that the blood gets finely and uniformly coated over millions of particles of dry R.C.

The polythene sheet over which the ingredients are mixed serves a good purpose; it prevents the seepage of the blood into the cemented platform. For the same reason, the initial drying of the mix should be carried out on the sheet itself. Once partially dry, the mix could be transferred onto the platform for it to dry completely.

Many factors such as prevailing temperature, wind velocity, relative humidity, size and nature of the absorbent, ratio of the absorbent to the blood, frequency of raking the mix during the period of its drying, etc. influence the rate of drying.

If the ratio in which blood and absorbent have been mixed is known and protein content of the absorbent is also known, it is possible to approximately calculate the protein content of the final dry mix. All these aspects have been discussed in detail in the literature.[1,2]

It is possible to make use of an absorbent for more than once; in other words, an absorbent once coated with blood and dried, could be reused again as an absorbent. Hence depending upon the number of times the absorbent has been used for the purpose, the resultant product could end up with a very high protein content. For example, if the protein content of an absorbent is 10 per cent, and it has been mixed with blood in the ratio of 1:1.5 (W/W), the dry product will end up with a protein content of approximately 27 per cent – an increase of about 17 per cent over the absorbent in the first mixing itself.

5.4 Sterilization

The absorbent + blood mix (dried) as discussed, should preferably be sterilized before use in livestock feeds. This can be done using semi-moist rendering vessel, the functioning of which has been described in chapter 3. Instead of meat, the material to be sterilized is packed in the space over the circular plate D and the equipment operated as discussed for rendering of meat. So treated material is found to be sterile.

5.5 Nutritional Value of Blood

Proteins are composed of amino acids. Certain of these amino acids are considered essential in as much as they can not be synthesized by the body. Nutritionally, quality of a protein is related to its ability to supply essential amino acids in the amounts needed. The amino acids have to be biologically available as to be absorbed and utilized by the body.

Human beings cannot synthesize 8 of the 23 amino acids of which proteins are composed. Meat not only contains a high percentage of the essential amino acids but these are present in such proportions that nutritional requirements of humans can easily be met. Hence meat protein is considered of high quality. A comparison of essential amino acids available in meat (beef) and cattle blood is shown in Table 5.2.

Whole blood protein, as can be seen, contains all the essential amino acids but the percentage of isoleucine and methionine is low compared to meat. As such, nutritionally, blood alone cannot adequately supply all the essential amino acids required in human diet. At the same time, blood is an excellent source of lysine and leucine – higher than available in meat.

Table 5.2 : Essential Amino Acids in Beef and Cattle Blood

Amino acid	Percentage of essential amino acid in	
	Beef	Beef Blood
Isoleucine	5.1	0.4
Leucine	8.4	13.6
Lysine	8.4	9.4
Methionine	2.3	1.8
Phenylalanine	4.0	8.0
Threonine	4.0	8.0
Tryptophan	1.1	1.4
Valine	5.7	8.0

Vegetable proteins are deficient in one or more of the essential amino acids, including lysine. Hence blood can be used to complement vegetable proteins as to evolve nutritionally more balanced feeds. It is for this reason that blood has been used in livestock feeds.

References

1. Mahendra Kumar (1989) : *Handbook of rural technology for the processing of animal byproducts (FAO Agricultural Services Bulletin No. 79)*. Food and Agriculture Organisation of the United Nations, Rome.

2. Mahendra Kumar (1987) : *Processing of animal byproducts in developing countries – a Manual*. Commonwealth Science Council, Marlborough House, Pall Mall, London, SW1Y 5 HX.

3. Divakaran, S (1982) : *Animal blood – processing and utilization*. Food and Agriculture Organisation of the United Nations, Rome.

4. Divakaran, S (1980) : *Animal blood in Food, feed, fertilizer, industry and laboratory*. NICLAI publication – Central Leather Research Institute, Chennai-600020 (India).

5. Rao, N.M (2001): *Processing of blood, in "Technologies for value realization of carcass byproducts in developing countries – a handbook"* edited by Ramasami, T., Ranganayaki, M.D. and Rajagopal, N.R. Central Leather Research Institute, Chennai-600020 (India).

6. Horace Thornton (1957) : *Textbook of meat inspection (3rd ed.)* Bailliere, Tindall and Cox, 7 and 8 Henrietta Street, London WC2.

7. Mann, I (1962) : *Processing and utilization of animal byproducts (FAO Agricultural Development Paper No.75)*. Food and Agriculture Organisation of the United Nations, Rome.

8. Mann, I (1984) : *Guidelines on small slaughterhouses and meat hygiene for developing countries (VPH/83.56)*. World Health Organisation, Geneva.

9. West, J (1974) : *Chemical preservation of blood, MIRINZ No.411*. The Meat Industry Research Institute of New Zealand, Hamilton.

CHAPTER 6

RUMEN CONTENTS

6.1 Definition

Rumen contents (R.C.) consist of partially digested feed materials consumed by the ruminants, together with digestive juices and millions of bacteria found in the rumen. R.C. is also known as ruminal contents (R.C.) or rumen digesta (R.D.).

6.2 Rumen – A Living Factory

The rumen, also called the "first stomach" or the "paunch" is a large sack wherein the animal is able to accumulate large quantities of feed and fodder. This apart, rumen is an anaerobic chamber and virtually a factory where several processes take place. The fodder is mechanically broken, mixed with digestive juices and cellulose is split by innumerable number of microbes belonging to anaerobic and facultative bacteria, various species of protozoa and fungi present in the rumen. The bacteria, in turn, is consumed by the animal as a rich source of protein. The vitamins already present in the fodder may be preserved and others, particularly vitamin B complex, are synthesized. It is therefore obvious that the rumen digesta is much higher in proteins and vitamins compared to the fodder on which the animal has been maintained upon.[1,2,3,4]

6.3 R.C. as a Useful Raw Material

R.C. finds many uses, the chief ones being for the production of

1. Biogas
2. Livestock feed
3. Compost/Vermicompost

Production of biogas, compost and vermicompost from R.C. have been discussed in Chapters 9 and 10 respectively. Consequently, production of livestock feed only will be discussed in this chapter.

6.4 Composition

The composition of R.C. depends upon the species of the animal as well as the type of feed on which it has been maintained upon. Analytical results of 4 types of samples are given in Table 6.1.

Table 6.1 : Analytical data of four types of R.C.

S.No.	Item	R.C.from sheep[5]	R.C. from lamb[6]	R.C. from cattle maintained on poor grasses[2,3]	R.C. from cattle maintained on green grasses[2,3]
1.	Moisture	8.51	—	9.32	8.90
2.	Ether extract	2.69	4.60	1.50	3.34
3.	Crude protein	17.51	28.80	11.01	17.20
4.	Crude fibre	32.76	25.50	32.45	24.83
5.	Phosphorous	0.80	0.56	0.65	0.83
6.	Calcium	1.60	1.00	0.99	1.32

The crude protein content of the poor grass on which the cattle was maintained as indicated in the table above, was below 3 per cent but that of the resulting R.C. being 11.01 per cent; this happens because the ruminants are able to convert the available nitrogen in the feed into protein on the one hand and the presence of a very large population of micro flora in the rumen on the other. The protein content in the other three samples viz. sheep, lamb and cattle (maintained on green grasses) is 17.51, 28.80 and 17.20 respectively. That goes to show that protein content by R.C. even from animals maintained on poor grasses is beyond 10 per cent and could reach almost 30 per cent in some of the ruminents. The fibre content is relatively high but all samples contained some fat, calcium and phosphorous; other minerals like Copper, Manganese, Zinc, Iron, Magnesium, Potassium and Sodium in macro and micro quantities have also been reported to be present.

Presumably, the quantity as well as the profile of the minerals available should depend upon the types of grasses on which the animal has been maintained upon; consequently, the profile of R.C.

from goats which have a habit to graze upon varied types of herbs and grasses is likely to be considerably different than the one obtained from a cattle. This apart, rumen contents contain not only the vitamins in the feed ingested before slaughter but also B vitamins from the rumen flora.[6] Putting all these factors together, R.C. could be legitimately considered as a good candidate for use as a feed material – for ruminants in particular because higher fibre content in their feed does not pose any problem.

6.5 Yield

The weight of wet R.C. as a percentage of live weight of the animal is roughly around 10-15 per cent. The dry matter as a percentage of wet weight of R.C. varies between 12-20 per cent; the same being comparatively lower for R.C. obtained from animals maintained on green and soft grasses compared to the ones consuming dry grasses. On an average, a fully grown cattle yields about 4 Kg of dry R.C.; the corresponding quantity for sheep and goat being roughly 0.4 Kg.[2,3]

6.6 Preparation of R.C. for Livestock Feed

R.C. from inspected and slaughtered stock alone should be considered for livestock feeds, that from dead stock and sick animals should be diverted for the production of compost/vermi-compost and methane.

There are three simple methods of converting R.C. into livestock feed, which being

1. Sun drying

2. Ensiling

3. Acid preservation

All the three techniques are simple requiring minimal equipment and will be discussed in detail.

6.6.1 Sun Drying

After the animal is slaughtered, the contents from all the four stomachs, namely the paunch (Rumen), honey comb (Reticulum), bible bag (Omasum) and rennet (Abomasum) are collected in their entirety. All the four stomachs are cut open with a butcher knife and contents taken out. The contents from cattle, sheep and goat could all be pooled together and mixed well before drying. If the contents are too watery, the same could be discarded. When the beast is starved before its slaughter, the stomach contents become watery; the longer the period of starvation, the more watery the contents are.

Drying is done by spreading the contents on a cemented platform in the open sun (Fig. 6.1)

Fig. 6.1 : R.C. being dried in the open sun.

The spreading should be done in a layer of about 6-8 cm thickness. If the layer is too thick, fermentation at the bottom of the layer may set in. Drying can also be done on iron sheets, metallic trays or granite slabs. In fact, any surface which absorbs heat at a fast rate is ideal to be used as a base for drying of the contents. When such base materials are not available, polythene sheets or mats may be used. However, period of drying on polythene sheet or a mat will be longer, when compared to the hard surface like cemented platform.

The contents should be raked every now and then while being dried. Raking helps in quicker and uniform drying of the contents; more often the raking, quicker the drying.

The contents have a tendency to form lumps while drying. When semi dry, the lumps could be opened up with ease if beaten lightly with the rake. The process of drying has to be continued till the contents are completely dry. Pounding of the dry product with pestle and mortar helps in opening the lumps further. For best results, however, the dry product should be milled in a hammer mill. Normally, however, these procedures are not required.

The time required for drying the rumen contents is dependent upon the prevailing temperature, relative humidity and wind velocity. Higher the temperature and wind velocity, shorter the time for drying. Too strong wind, however, is not desirable as it carries away the dry and semi dry material with it. Higher the relative humidity, lower the rate of drying and vice versa. More details about sun drying of R.C. and blood loaded on dry R.C. are explained in the chapter on "Blood".

The dry contents may have to be sterilized before it is used in livestock feeds, even though it has not been found necessary when the R.C. have been obtained from healthy and inspected animals. This is best done by steaming of the same; the semi-moist rendering vessel discussed in Chapter 3, could be used for the purpose of sterilization.

6.6.2 Ensiling

Ensiling is a process of keeping any green plant material air tight, wherein it ferments anaerobically (in the absence of air); the fermented product formed is called Silage and the container in which the process is carried out is known Silo. Major changes during ensiling are fermentation of carbohydrates to form acids such as lactic acid, acetic acid and breakdown of some forage proteins to simpler compounds including ammonia. The fermentation process occurs during the first one and a half months; afterwards the silage remains practically unchanged for 12-18 months.[7] This technology was first developed for the preservation of plant material and subsequently adopted for ensiling animal wastes as well.

Szember[8] prepared silage from R.C. after addition of 5 per cent molasses. Rao and Fontenot[9] studied the use of R.C. and whole bood in the ratios 2:1 and 1:1 (w/w) with ground wheat straw in the proportions of 70:30, 60:40, 50:50 and 40:60 (w/w) with and without dry molasses (5% w/w). The mixtures were ensiled for six weeks in polyethylene bags. It was observed that all silages had desirable aroma after 6 weeks of formualtion, pH ranging between

4 and 5 and lactic and acetic acid levels were sufficient to ensure preservation in most of the ensiled mixtures.

Unlike sun drying which is dependent on weather, ensiling could be practiced under all climatic conditions.

The technique is simple amounting to mixing of the R.C. with some other ingredients and packing the mix tightly in silos or other suitable containers.

The dry matter content of the material to be ensiled may vary between 35 to 55 per cent. As discussed earlier, the dry matter content of raw R.C. varies between 12 to 20 per cent; the same should be adjusted before ensiling.

There are two simple ways of adjusting the moisture content. The R.C. could either be mixed with a variety of dry crop residues (which have suitably been grounded in a mill, such as paddy straw, wheat straw, corn stalks, and the like) or by partially drying of raw R.C. in the direct sun on a cemented platform. On a clear sunny day, it is possible to bring down the moisture content of the R.C. within desirable limits required for ensiling by the afternoon. Accordingly, the following three compositions were prepared; all components were mixed thoroughly before ensiling.

1. a. R.C.   40 parts by weight
    b. Blood   20 parts by weight
    c. Paddy straw (ground)   40 parts by weight
    d. Molasses (60% solids - min)   5 parts by weight
        – Moisture content of the above mix = 53.4 per cent[10]

2. a. R.C. partially sun dried   100 parts by weight
        – Moisture content of the above R.C. = 52.5 per cent[11]

3. a. R.C. Partially sun dried   100 parts by weight
        as at (2)
    b. Molasses (60% solids - min)   5 parts by weight
        – Moisture content of the above mix = 53.2 per cent[11]

The three treatments were ensiled by filling them tightly in separate cylindrical metallic drums, double lined with polythene bags. After the materials have been packed, the open ends of the polythene bags were tied tightly as to create anaerobic conditions (Fig. 6.2). The drums were finally closed with lids. The ensiling period was 6 weeks, after which the drums were opened to draw samples for analysis.

Fig. 6.2: The mix to be ensiled, packed tightly inside the drum double lined with polythene bags.

All the three treatments were analysed for various parameters such as moisture, solids, proteins, carbohydrates, pH, bacteriological status, etc., before and after ensiling. From the original pH of around 7 for all the three treatments, the same dropped down to around 4.5 to 5 after ensiling; all the three post ensiled mixtures were free from *E. coli* and *Salmonella*. All the three ensiled products developed good aroma and had free flowing texture.

During the period of ensiling, the soluble carbohydrates present in the system are converted into acids such as lactic and acetic – lactic acid being predominant.[12,13] This is caused by the lactic acid bacteria already present in the system and is responsible for the drop of the pH from the original 7 to around 4.5 to 5. Lowering of pH and the anaerobic conditions are responsible for the elimination of *E. coli* and *Salmonella* from the system, resulting in a product which can be safely fed to the animal. The protein content of the product also remains intact with the decrease of only a very small percentage of the same during the process of ensiling.[10]

6.6.3 Acid Preservation

As the name indicates, this is more of a preservation technique for the conservation of the R.C. Once preserved, the same keeps without problem for long periods.

The process consists in mixing of the R.C. with sulphuric acid as to lower the pH to 4.[6] Sulphuric acid should be handled with care as improper handling can cause problems. The acid should also be neutralised before any further processing or use of acid preserved R.C. is planned.

6.7 Profile of Coarse and Fine Components of R.C.

Most of the proteins and soluble carbohydrates are present in the liquid portion of the R.C. as obtained after slaughter of the animal.[14] When dried, the liquid part becomes more powdery than the fibrous part. The author sieved several samples of dried cattle rumen contents, into its fine and coarse components; the sieve having 150 holes/cm^2. Both the components were weighted; the finer component varied between 40-60 per cent – the average being 50 per cent of the total dry weight of the R.C. The two components were analysed for crude protein and crude fibre – the results are presented in Table 6.2.

This is a very simple method of separating rich protein component of the R.C. which could preferentially be used in fish and poultry feeds. This aspect will be discussed under 6.8.

Table 6.2 : Analytical profile of fine and coarse components of R.C.

Item	Crude protein *	Crude fibre*
Fine component	19.2 per cent	20.3 per cent
Coarse component	6.4 per cent	39.5 per cent
Original without sieving	12.6 per cent	29.7 per cent

* Average of 6 samples; all results on moisture free basis.

6.8 Use

Sun dried R.C. should be sterilized before use as discussed under 6.6.1. However the same has been reported to have been used safely even without sterilization in poultry, pig and ruminant feeds, in case the material has been obtained from inspected and disease free animals. In fact, the use of sun dried product in calf starter rations is even desirable. The natural bacterial flora present in R.C. does not get completely destroyed during sun drying of the same. When mixed with calf starter rations, it helps in early establishment of normal ruminal flora in young calves.[1,2,3] In the case of poultry and ruminant feeds, R.C. has been reported to have been used upto 10 per cent and 40 per cent respectively. In the case of pig diets, R.C. has been used to replace 100 g by weight of grain without a decrease in weight gains or feed efficiency.[6]

As is clear from Table 6.1, the fibre content of R.C. is high and this limits its use in poultry and fish feeds; poultry and fish can not tolerate high fibre content in diet. Ruminants, however, can tolerate high fibre feed. For this reason, the fine and coarse components of R.C. as discussed under 6.7 could be advantageously used in poultry/fish and ruminant rations respectively.

Ensiled R.C. has been reported to have been used satisfactorily as a feed for cattle, buffalo and sheep. Logically, it should be possible to make use of it in diets of those monogastric animals which tolerate high fibre content such as mule and horse. Rumen content silage is reported to be palatable to pigs, which can consume upto 0.5 kg per day once they become accustomed to it.[6]

R.C. is a fermented product and contains growth factors, especially the B complex vitamins and some essential amino acids possibly derived from micro flora. This apart, as has been discussed under 6.4, R.C. has fairly high crude protein content, contains some fat and is rich in a variety of essential mineral elements. Because of these factors, R.C. is ought to find greater use in animal diets.

6.9 Relation Between Volume and Weight

An idea of volume to weight ratio of the wet, dry and ensiled R.C. will help in calculating the storing space required for the respective commodities. The data given below will be of use in this connection.

1. Wet R.C.   – 1 m^3 = 750-800 kg.

or

1 lit = 0.75 to 0.80 kg.

2. Dry R.C.   – 1 m^3 = 150-160 Kg.

or

1 lit = 0.15 to 0.16 kg.

3. Ensiled R.C.   – 1 m^3 = 650-700 kg.

or

1 lit = 0.65 to 0.70 kg.

6.10 Space for Sun Drying

A cement platform measuring around 10 m × 10 m is sufficient for drying R.C. obtained from 7-8 cattle a day.

6.11 Storage

Sun dried R.C. should be stored in a dry place. Damp place is not suitable. The moisture content of the product should not exceed more than 10 per cent.

The ensiled product should preferably continue to remain under anaerobic condition until required for use.

One has also to be careful about the handling of the produce such that it does not get contaminated at any stage of handling. Hands and handling tools have to be clean and one has to be extra careful that it does not get contaminated by shoes either.

References

1. Mann, I (1962) : *Processing and utilization of animal byproducts (FAO Agricultural Development Paper No. 75).* Food and Agriculture Organisation of the United Nations, Rome.

2. Mahendra Kumar (1987) : *Processing of animal byproducts in developing countres – a Manual.* Commonwealth Science Council, Marlborough House, Pall Mall, London, SW1Y 5HX.

3. Mahendra Kumar (1989) : *Handbook of rural technology for the processing of animal byproducts (FAO Agricultural Services Bulletin No.79).* Food and Agriculture Organisation of the United Nations, Rome.

4. Rao, N.M. and Mahadeswaraswamy (2001): *Processing of rumen contents in "Technologies for value realization of carcass byproducts in developing countries – a handbook"* edited by Ramasami, T., Ranganayaki, M.D. and Rajagopal, N.R. Central Leather Research Institute, Chennai-600020 (India).

5. Sastry, T.P., Rao, N.M. and Scaria, K.J (1983) : *Studies on the proximal analysis of ruminal contents.* Leather Science, 30(5), 145-48.

6. Gohl, B (1981) : *Tropical feeds (FAO Animal Production and Health Series No.12, Page 396).* Food & Agriculture Organisation of the United Nations, Rome.

7. Wilkins, R.J. and Wilson, R.F (1970). J.Br. Grassland Soc. 26, 107.

8. Szember, A. (1968). *Development of some groups of micro-organisms during ensiling of rumen contents. Nutr. Abstr. and Rev.* 36, 6653 (Abstr.)

9. Rao, N.M. and Fontenot, T.P (1987): Anim. Feed Sci. and Tech. 18, 67-73.

10. Rao, N.M., Sastry, T, P., Rose, C., Biswas, G and Mahendra Kumar (1990) : *Studies on silage fermentation in certain slaughterhouse byproducts.* Proceedings of sixth international symposium on agricultural and food processing wastes held at Chicago between 17 and 18 Dec.

11. Mahendra Kumar (1990) : *Unpublished data.*

12. Lindergren, S (1983) : *Some aspects of the use of microbial cultures of fermentation and storage of feed products.* Thesis, Swedish University of Agriculture Sciences, Department of Microbiology Uppsala.

13. Lockwood, L.B (1979) : *Production of organic acids by fermentation in Microbial Technology, Vol. 1,* pp 355-387, Academic Press, New York.

14. Fernando, T (1980) : *Utilisation of paunch content material by ultrafiltration.* Process Biochemistry, 15 (3), 7.

Chapter 7

Minor Products

7.1 Introduction

Under this chapter, some of the raw materials whose share as a percentage of the total weight of the animal is small but may be important from the point of view of their utility for the production of useful end products, will be discussed in brief; such raw material being glands and organs, intestines, horns and hoofs, hair and gall baldders.

7.2 Glands and Organs

A variety of enzymes, harmones, biomedicals which are not only of high value but some of them even life saving are prepared from a number of glands and organs. They find innumerable applications in food, pharmaceuticals, chemical, agriculture, therapeutic, diagnostic, molecular biology, health, animal breeding, artificial insemination etc. some of the glands and organs widely used for the purpose are listed in Table 7.1

Table 7.1 : Glands and Organs

Glands :	Pancreas, Thyroids, Parathyroids, Suprarenal gland or Adrenal bodies, Ovaries, Testes, Thymus, Pituitary body
Organs :	Liver, Stomach, Placenta, Gall bladder, Intestinal mucosa.

While this is so, their manufacture should be attempted with utmost care and planning. This should only be taken up under most hygienic conditions and when the number of healthy and properly inspected animals slaughtered under one roof is quite large as to get economically viable quantities of raw material for manufacturing. Table 7.2 provides an idea of weight of glands and organs per animal head. Expected revenue returns from slaughter of 1000 animals is given in Table 7.3.

Table 7.2 : Major Products and their Contents in various Organs and Glands

Source material	Animal	Weight of Raw material per animal (average)	Products	Weight of Product per unit raw mat.
Pituitary	cattle	3g	LH	80mg/kg
			FSH	3 mg/kg
			PRL	150mg/kg
	sheep	0.6g	LH	60mg/kg
			FSH	200mg/kg
Adrenal	cattle	20g	Adrenalin	13mg/g
Thyroid	cattle	30g	Thyroxine	40mg/kg
	sheep	5g	Thyroxine	35mg/kg
Thymus	calf	80g	Histone	1.5g/100g
			Enzymes	3.5g/kg
Liver	cattle	4kg	Catalase	0.8g/kg
Testis	cattle	300g	Hyaluronidase	0.6g/kg
	sheep	70g	Hyaluronidase	0.75g/kg
Stomach	calf	60g	Rennin	0.1g/kg
(mucosa)	cattle	250g	Repsin	0.175g/kg
Intestinal	calf	70ml	Alkaline	0.05g/lit
Mucosa			Phosphatase	

Table 7.3 : Returns on End Products from Slaughter of 1000 Animals

End Product	Animal	Gland Organ	Total of Raw material	Total Qty. of end product	Expected value in US $
LH	Cattle	Pituitary	3.0 kg	0.24 g	28800
	Sheep	Pituitary	0.6 kg	0.036 g	2160
FSH	Cattle	Pituitary	3.0 kg	0.009 g	9000
	Sheep	Pituitary	0.6 kg	0.12 g	1200
PRL	Sheep	Pituitary	3.0 kg	0.45 g	1500
Thyroxine	Cattle	Thyroid	30.0 kg	1.20 kg	688
Insulin	Cattle	Pancreas	300.0 kg	0.045 kg	2556
Pancreatic Enzymes	Cattle	Pancreas	300.0 kg	1.05 kg	4069
Catalase	Cattle	Liver	4000.0 kg	3.2 kg	876
Hyaluro-nidase	Cattle	Testis	300.0 kg	0.18 kg	236

Such large abattoirs and requisite hygiene are rarely available in most of the developing world. Consequently discussion on glands and organs falls outside the perview of this book. The discussion here will be restricted only to four raw materials viz. intestines, horns and hoofs, hair and gall bladders. The collection and preservation of these raw materials does not demand very critical conditions and their subsequent processing into products is comparatively easy. However, readers interested in glands and organs will find a brief account both on collection of raw material and processing and use of end products in the literature.[3]

7.3 Intestines

Intestines of herbivorous animals like cattle, buffalo, sheep and goat and that of pig have been used for a number of purposes. Being tubular in construction and that of animal origin, the intestine after proper processing, has been very commonly used as a container or a "casing" for an edible meat product "sausage". Intestine apart, other parts of alimentary tract have also been used as sausage casings; this aspect has already been briefly discussed in Chapter 2.

The other uses of intestine had been in the manufacture of round transmission belts, sports guts, musical guts and surgical catgut. Selected membraneous parts of intestine have been used as dialysing membrane and for the production of a product called "collagen sheet" which has been found to be extremely useful as a preliminary dressing material in the management of all types of open skin surfaces such as burns, traumatic wounds, ulcers, amputated surfaces and so on. Processes for both surgical catgut and collagen sheet were developed in India and are now commercial under trade names "Integrin" and "Köllagen" respectively. Köllagen is the only product of its kind available in the market.

The processing details of intestines into casings, their preservation, grading, standards, quality control, packing and information on the manufacture of products like belts, sports guts, musical guts,[2,3] surgical catgut, collagen sheet and test procedures etc. of these products[4,5] are available in the literature. It is outside the scope of this write up to deal with these details and one is advised to refer to the literature for detailed information on respective products.

7.4 Horns and Hoofs

Horns and hoofs are made up of a protein called "Keratin". Other than horns and hoofs, hair, wool, nail, quill and feathers are examples of materials composed mainly of the protein "Keratin". Keratin is characterised from other proteins by the presence of high percentage of sulphur containing amino acid "cystine". The percentage of sulphur present in keratins, mostly contributed by cystine, varies between 1-5 per cent. For this reason, keratins having higher percentage of cystine have been used for the maufacture of the same (cystine). Cystine finds use in bakery and other food products and sells at attractive prices; present international price being around US $30/kg.

Mention of the preparation of fancy articles and horn and hoof meal from horns and hoofs has already been made in Chapter 2. Horns and hoofs have also been used for the preparation of foam compounds, soup flavours and many more products.

Horn and hoof meal is very rich in nitrogen and is extensively used as an organic fertilizer.

Foam compounds made out of horns & hoofs are specially useful for extinguishing fire of solvents. When sprayed, the foam settles down on the surface of the solvent and thus cuts down the supply of air to the burning solvent. This results in the termination of fire. Water is heavier than solvents. When water alone is sprayed, the same settles down underneath the solvent and hence does not help in extinguishing the fire.

A number of meaty flavours are made from horns and hoofs by the so called "acid hydrolysis process". These flavours have been used in soups to develop strong meat taste and flavour.

Methods of processing horns and hoofs into products like horn and hoof meal,[2] foam compound[6,7] and soup flavours are surprisingly simple and details are available in the literature

7.5 Hair

As already discussed under horns & hoofs, hair is also made up of the protein "Keratin". Keratins are very resistant to the action of water, organic solvents and mild acid solutions. For this reason, hair is ideally suited for the manufacture of brushes particularly for paints based on solvents. Mention about the use of hair for the manufacture of brushes has already been made in Chapter 2. Hair also finds a number of other uses – upholstery filling, insulation, carpet under felts, textiles, druggets and ropes. However, the use of hair for brush making ensures the maximum economic return.

Before the hair could be utilized for brush making, the same is required to be properly collected, conserved and dressed to a state that it is ready for making brushes. Dressing in itself consists of a number of intricate operations requiring careful handling of the material.

Detailed information on the procedures of collection, conservation and dressing of hair as well as manufacture of certain types of brushes is available and one is advised to refer to the literature for more information.[2,3]

7.6 Gall bladder

Gall bladder is a membraneous bag, attached to the liver, and is meant to store bile, the mention of which has already been made in Chapter 2. This apart, gall stones may also be available from a small number of bladders.

As discussed under Chapter 2, bile is a syrupy liquid, dark yellow to green in colour; a cattle may yield upto 500 cc of the liquid, the corresponding yield from a sheep and goat being upto 50 cc.

Bile is a very good emulsifying agent and infact, it is utilized by the animal for the emulsification of fats and oils present in the feed. For the same reason, bile has been found to be useful for the cleaning of hair, butcher knives and the like.[2,3]

Bile is a very important raw material for the manufacture of a number of products like cholic acid, desoxy cholic acid, salts of these acids and cortisone for use in pharmaceuticals; as a result many countries import properly preserved bile for the manufacture of these products.[2]

Preservation of bile is very simple and is basically done by the evaporation of raw bile in a steam jacketed kettle to a concentration of about 75 per cent solids; in its original raw state, bile has around 8.5 per cent solids on an average, with a specific gravity of 1.025. Concentrated bile is also called "Inspissated Bile" and preserves well for long periods at ambient temperatures; full details for concentrating the raw bile liquid are available in the literature.[2]

Gall stones are available in varied colours from reddish yellow to yellow black and in various shapes – rounds, oval, triangular and polygonal. Bigger the size of the stones, higher the prices offered. Size to size, reddish yellow stones fetch higher prices. The present prices offered are above US $ 1000 per kg of good quality stones.

The use of gall stones is not fully known but they are said to be used as lucky stones or in medicines by the orients; Chinese and Japanese are the main buyers of gall stones. The collection of gall stones is rather simple; the contents of the gall are strained through a filter whereby the stones are retained on the filter. The same are collected and dried under shade; if dried in the sun, they have tendency to blacken.[2]

References

1. Gade, W.N (2001) : Processing of endocrine glands and organs, in *"Technologies for value realization of carcass byproducts in developing countries – a handbook"* edited by Ramasami, T., Ranganayaki, M.D. and Rajagopal, N.R. Central Leather Research Institute, Chennai-600020 (India).

2. Mahendra Kumar (1989) : *Handbook of rural technology for processing of animal byproducts (FAO Agricultural Services Bulletin No.79).* Food and Agriculture Organisation of the United Nations, Rome.

3. Mahendra Kumar (1987) : *Processing of animal byproducts in developing countries – a manual.* Commonwealth Science Council, Marlborough House, Pall Mall, London SW1Y 5HX.

4. Barat, S.K. and Mahendra Kumar (1959) : A process for the utilization of mammalian intestines for the manufacture of absorbable surgical catgut/suture/ligature "Plain and Chromic" or the like. Ind. Pat. No. 116762, 3rd Dec., 1959.

5. Mahendra Kumar (1984) : A process for the production of collagen sheet material from mammalian tissues. Ind. Pat. No. 154280, 13th Oct., 1984.

6. Mitsubisi Chemical Industries Co. Ltd. (1980) : Fire extinguishing foams from Keratin hydrolysate. Jpn. Kokai Tokyo Koho 8, 070, 272, May, 27 (1980).

7. Hoshino, M (1980) : Keratin hydrolysate foaming properties for fire extinguishers. Yakagaku 29(2), 102 (1980).

CHAPTER 8

SANITATION AND HYGIENE

8.1 General

We live in an ocean of agglomerate of dirt, dust and microbes all around us, it (agglomerate) is there on every conceivable object– our body, bed, pillow, kitchen, food and what not; think of anything and it is surely present there.

In the context of slaughtering and rendering, the contamination of meat and feed with the agglomerate may cause serious health problems both to humans and animals. The contamination should, therefore be minimized by proper measures taken in hygiene and sanitation. The two words *viz.*, hygiene and sanitation are inter-related in as much as sanitation is equated with conditions which offer pubic health while hygiene means sanitary science or ways and means to advance sanitation.

Consequently, sanitation in the slaughtering and rendering industries encomprises all environmental elements which are likely to affect public health and hygiene and will embrace any conceivable intervention likely to help in safeguarding public health from causative factors.

With the above introduction in view, some of the issues involved are discussed. The discussion at best can be considered as suggestive and indicative; it is left to the discretion and imagination of all concerned to improve upon and go a mile further towards the set goal of hygiene and sanitation.

8.2 Building

There are a few essential requirements which should work as guideline while constructing either a slaughtering or rendering facility. The building should be constructed on a piece of elevated land to avoid possible flooding during rains; marshy and rocky land should be avoided. The location has to be well away from villages, towns and cities.

The building has to be well ventilated, flyproof, having smooth floors and walls and with sufficient work place with ease without any hindrance. It has also to be well drained to avoid any accumulation of water. Constructed area will comprise of office rooms, toilets with bath facility for changing dress for workers, shed for vehicles, godowns for processed products, green area for plants for healthy environment, suitable effluent treatment facility (different designs have been offered under chapter 12), and clean and well kept surroundings. The dirty and clean areas of the building have to be well segregated to minimize contamination of the produce. Entry for vehicles, staff and visitors should be well planned such that it is away from the production area.

8.3 Water

There has to be abundant supply of clean potable water. Water is required everywhere–in flaying yards, toilets, personal hygiene, watering of plants and so on. Ideally, it should be stored in an overhead tank so that it is available with force at the point of delivery; it reduces work and helps in better and faster cleaning of the place or for that matter of watering the plants.

However, in the context of developing countries, the availability of electricity for pumping up water in overhead tank is either erratic or not even available in some place. In such situations, water hand pumps have to be installed at the highest elevated place possible and water stored in covered tanks for smooth running and cleaning of the place. Suffice it is to say that sanitation and hygiene are of utmost importance in slaughtering and rendering establishments, which to a good degree is dependent on abundant availability and supply of potable water.

8.4 Transport Vehicles

Depending upon the load and volume of goods to be transported, all types of transporation means like truck, van, jeep and bullock/horse driven carts are in use. However, mechanised transport has been found to be better. As nature of materials to be transported is highly perishable (meat, dead stock, byproducts etc.), a mechanized vehicle will ensure faster transportation which is so desirable.

The vehicles should be kept clean, tidy and in good working condition to minimize chances of contamination in the work place. The vehicles have to be regularly washed (preferably with hot water and steam), disinfected and completely dried before use. The hygiene of the personnel working in transportation has also to be ensured.

8.5 Means of Communication

With the fast development of communication technology with the availability of cell phone, communication has become very easy and dependable. However, cell phone is not either available at affordable price not the connectivity is yet reliable in many countries of the developing world–rural areas in particular. To overcome this problem, a very practical and innovative system of communication has been evolved in India for the collection of dead stock. The establishments processing fallen stock have appointed informers in all the villages in the area of their jurisdiction; soon after the death occurs, the concerned informer in the village travels on bicycle to the establishment and informs about the death of the animal. After receipt of information, the establishment organizes the collection and transport of the dead stock as fast as possible. Under field conditions as available in India, most of the carcasses usually get collected and transported to the establishment within a period of about six hours after death. This has been further cut down significantly wherever cell phones have been introduced. As no significant damage occurs to the carcass within 12–15 hours after death, this method of communication has been found quite satisfactory and yet simple and cheap. Dependable and faster communication prevents decay of the dead stock, resulting in the production of quality products and helps in achieving hygiene and sanitation.

8.6 Dry Slaughtering

There has to be an attempt to ideally eliminate or minimize contamination, both of the raw material as well as produce with dirt, dust and microbes from the environment or during handling. To this end, the flaying of the carcass should be done in a manner which is followed under "dry slaughtering" wherein the dressing of the carcass *viz.,* its flaying, evisceration, splitting and dispatch of the produce etc. is done in a manner that the flayed carcass does not come in contact with water, hide or viscera etc.[1] Care should also be taken that evisceration is done with utmost care so that bowels do not get punctured and contaminate the flayed carcass. These measures will ensure least possible contamination during dressing and handling operations.

Incidentally, dry slaughtering can be followed with comparative ease where number of animals to be slaughtered is not too large and hence ideally suitable for adoption in small slaughterhouses as well as for dead stock.

There are many other added advantages of dry slaughtering. The use of water during dressing operation is cut down significantly.[2] This curtails on the quantum of effluents generated. Detailed discussion on this aspect is available in Chapter 12.

From economic considerations also, dry slaughtering is desirable as it reduces use of water and cuts down on effluents as well as demand for land and infrastructure needed for treatment of effluents comes down significantly.

8.7 Plant Maintenance

It is important that plant and equipment are cleaned and disinfected at regular intervals and kept in proper working order; defects, if any, cause delays in execution of work. Additionally, uncleaned and rusty equipment becomes a source of contamination.

8.8 Education Programme

This phase should clearly specify as to what is to be done, how it is to be done and who is to do it with proper delegation of authority needed to carry out the job successfully and without delay.

Both workers and staff should be given lessons on the importance of maintaining hygiene and how to achieve it.

Workers should be trained in proper methods to carryout the job assigned. They should understand the importance of changing dress, wearing masks where necessary, washing hands with soap, their own personal hygiene and keeping tools and equipment clean.

They should also avoid trespassing from dirty and wet areas like flaying to dry and processed goods areas and vice versa.

8.9 Scheduled Diseases

Diseases like anthrax, foot and mouth, rabies, tetanus etc. fall under the category of scheduled diseases in as much as these are communicable to man. Among these, anthrax is the most dreaded one; it's spores pose great danger to man, livestock and environment with serious consequences. As such, animals having symptoms of anthrax or carcasses of such animals should not be used either for meat or rendering. These have to be discarded in their entirety and disposed off according to laid down procedures discussed in detail in Chapter 11. One has to choose from the available three methods *viz.*, incineration, burial or anaerobic digestion.

Unfortunately, there is no rapid spot test for anthrax. In areas where anthrax is not prevalent, the problem does not exist. However, in regions of incidence (of anthrax), clinical signs like higher fever, bloody discharge from natural body openings (like anus and nostrils) and sudden death (cattle in particular) could be the guiding factors. Such carcasses should be discarded and disposed off, even though some of them may not have died of anthrax.[3] Hence, the personnel responsible for inspection of the animal before slaughter or collection of the dead stock from the field should find the history of the animal as well as dead stock respectively with a view of disposal even in suspected cases.

As there are various forms of anthrax and different species of animals may show different clinical signs when affected, one is advised to know more about the disease.[4]

As a measure of safety and precaution, the personnel working with livestock should be vaccinated against anthrax, particularly in regions where anthrax is prevalent.

8.10 Microbes

The information reported under this *'head'* has been borrowed almost in it's entirety from the book *"Sanitation and Hygiene in the Production of Rendered Animal Byproducts"* with the permission of the author.[5]

Biology is defined as the study of living organisms and life processes, including their structure, function, growth, origin, evolution, and distribution. Microbiology is the branch of biology that studies organisms that are not directly visible to the unaided or naked eye called microorganisms. The following brief overview presents bacteria, molds, yeasts, and viruses and their potential impact on the slaughtering as well as rendering industries.

Bacteria, the group of microbes of greatest importance to slaughtering and rendering, are discussed to characterize their behavior, their relevance to sanitation and hygiene, and considerations for their control.

Bacteria (singular, bacterium) are small, single-celled organisms belonging to the world of plants, and universally distributed throughout our environment. Most bacteria are necessary and beneficial; some are destructive and harmful.

Bacteria are grouped into convenient categories according to such characteristics as shape, temperature preferences, nutrient needs, oxygen utilization, disease-producing potential (pathogenicity), microscopic characteristics, ecologic adaptations (environment in which organisms survive and reproduce).

8.10.1 Shape

Bacteria may be in 1 of 3 shapes:

1.   If they are spherical they are called cocci;
2.   If they are straight rods, they are called bacilli;
3.   If they are curved or spiral rods, they are called spirilla.

Cocci (Figure 8.1) may appear as separate cells in various groupings *e.g.*, in pairs–diplococci or in grape-like clusters *e.g.*, staphylococci and in chain-like structures *e.g.*, streptococci.

The cocci (diplococci, streptococci, and staphylococci) are among the most important disease-causing bacteria in animals and humans. For instance, they cause strep throat, tonsillitis, pneumonia, boils, staphylococcal food poisoning. Therefore, this group is of great clinical and public health significance.

The bacilli, or rod-shaped bacteria, contain many organisms of importance to agriculture, the slaughtering and rendering industries, and public health. They include bacteria responsible for meat spoilage, fermentation in milk and cheese, and diseases such as tuberculosis, anthrax, salmonellosis, and botulism–of great interest to renderers as well as meat and slaughtering establishements.

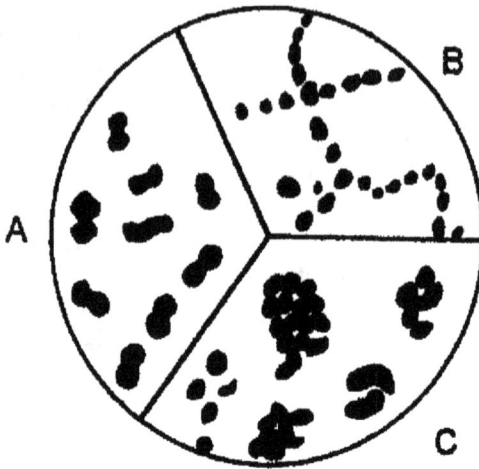

Fig. 8.1 : Cocci
(A) Diplococci; (B) Streptococci; (C) Staphylococci

8.10.2 Size

Microorganisms are measured in units known as microns. A micron is equal to 1/1,000 mm or 1/25,000 inch, and is abbreviated Gr mu. Members of the cocci group, *e.g.*, staphylococci, range from 0.5 to 1.2 mu in diameter.

8.10.3 Cell Structure

The bacterial cell is made up of a living material called cytoplasm which is surrounded by a thin cellulose membrane known as the cell envelope (wall). The cell wall is often surrounded by a jellylike material called slime. The growth of large numbers of bacteria results in concentration of this slime which can be seen on surfaces of contaminated equipment and spoiled meat. When the slime layer is thick and firm enough to have a form, it is referred to as a capsule (Figure 8.2).

The most important thing to remember about bacteria is that they are complex living organisms whose structural components have specific functions, similar to those of humans. Different kinds of bacteria have different requirements for oxygen and nutrients. They differ markedly with regard to the optimal temperature range for their growth. Some are essential for life. Others cause disease.

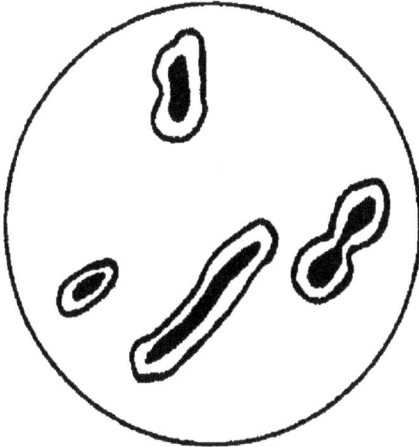

Fig. 8.2 : Bacteria within capsules

In essence, we are dealing with a diversity so unique that any effort to simplify the subject is impossible.

8.10.4 Motility

All particles suspended in air or liquid exhibit a spinning or oscillating motion designated Brownian movement. Bacteria possess this characteristic, but it is not considered their means of motility because it is caused by the bombardment by the surrounding water molecules. Many bacteria are equipped with flagella in the cell envelope. These threadlike appendages, composed entirely of protein, enable them to move about. The flagella are arranged in 3 patterns; monotrichous (single flagellum at one end), lophotrichous (tuft of flagellum at one or both ends), and peritrichous (flagella distributed over the entire cell) (Figure 8.3).

8.10.5 Multiplication and Growth of Bacteria

The population of microorganisms in the biosphere is roughly constant: growth is counterbalanced by death. Many microorganisms live in consortia formed by representatives of different genera. Growth is the orderly increase in the sum of all the components of an organism. Cell multiplication is a consequence of growth.

Fig. 8.3 : Bacteria with various types of flagella

Many bacteria reproduce by binary fission. The average time required for the population, or the biomass, to double is known as the generation time, or doubling time, in which the number of cells theoretically double.

Factors that inhibit maximum theoretical growth of bacteria include:

1. Overcrowding of cells
2. Outgrowing the food supply
3. Not all cells will produce
4. Acid production and other byproducts of cell growth could create an inhospitable environment for survival.

The growth of a bacterial culture follows a predictable pattern that can be demonstrated graphically as a growth curve (Figure 8.4).

The bacterial growth curve (Figure 8.4) is divided into 5 sections:

1. The lag phase (phase of adjustment)–represents the period during which the cells, depleted of metabolites and enzymes as the result of unfavorable conditions, adapt to their new environment. Little or no multiplication takes place at this time, there may even be a reduction in bacterial

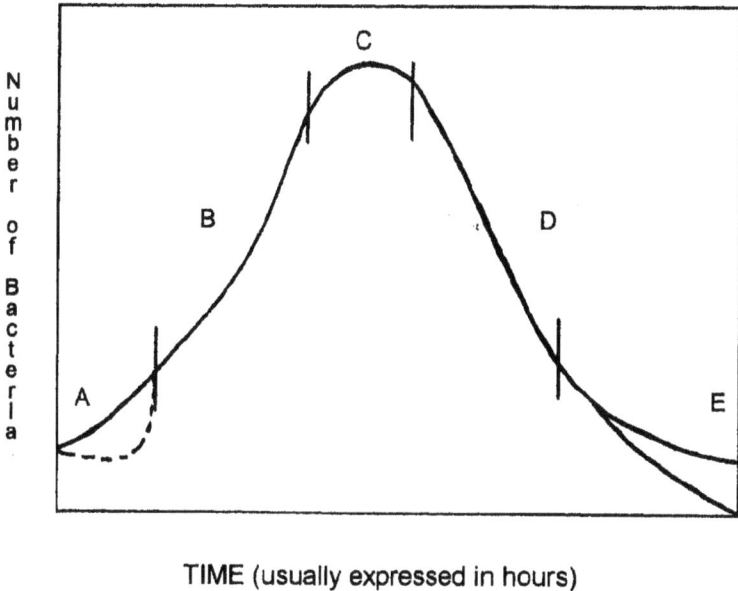

TIME (usually expressed in hours)

Fig. 8.4 : Bacterial growth curve at a constant temperature

numbers before the adjustment is made (represented by the dotted line). When enzymes and intermediates reach certain concentrations they permit growth to resume.

2. In the growth phase, bacteria undergo their maximum growth rate resulting in a dramatic increase in bacterial numbers. The packing house and rendering industries should be conscious of these phases because it is during this period that meat shows signs of spoilage; the same is true of protein meals exposed to moisture and stored at ambient temperature.

3. The resting phase is characterized by the exhaustion of nutrients or the accumulation of toxic products and the bacteria die as fast as they are being reproduced. There is a slow loss of cells through death, which is just balanced by the formation of new cells through growth and division.

4. The death phase is an acceleration of the resting phase in which the death rate increases until it reaches a steady

level. Frequently, after the majority of cells have died, the death rate decreases drastically, so that a small number of survivors may persist for months or even years.

5. In the readjustment phase, organisms may survive and become adjusted to their new environment.

8.10.6 Spore Formation

Members of several bacterial genera *e.g., Clostridium, Bacillus* are capable of forming spores as a response to unfavorable environmental conditions in which nutrients have become depleted. Each bacterial cell forms a single internal spore that is liberated when the mother cell undergoes autolysis (destruction). The spore is a resting cell that is highly resistant to heat, drying, and chemicals. After the return of favorable conditions, the spore germinates to produce a single vegetative cell. Spores are basically nothing more than the cytophasm of the vegetative cell in a condensed form and in this form they can remain dormant for long periods. Two diseases of interest to the slaughtering and rendering industries are anthrax and tetanus, and the causative organisms of both can form spore: *Bacillus anthracis* (anthrax spores) and *Clostridium tetani* (tetanus spores).

8.10.7 Environmental Factors

The environment plays a significant role in the types of bacteria present in any setting. Both slaughtering as well as rendering industries have to be fully aware of this responsibility. Rendering industry in particular, has to assume to change the microbial flora of incoming raw material from relatively contaminated to properly processed product that is "sterile". The various steps in the rendering process (cooking, time-temperature) inactivate organisms and control their growth. Thus, one of the rendering industry's important contributions is elimination of pathogenic organisms in material that could otherwise serve as reservoirs of disease transmission.

8.10.8 Nutrition

Bacteria and other microbes, as do all living organisms, have food requirements which the environment must provide if they are to live there. Different species of bacteria have nutrient requirements that are fairly species specific. Nonetheless, all bacteria require nitrogen, carbon, an energy source, trace minerals, and moisture. Slaughtering and rendering plants readily provide bacteria with ideal growing conditions. Thus, both industries must use innovative measures to reduce and eliminate factors that perpetuate growth and survival of organisms, for instance, by avoiding the use of water that would provide moisture for replication and growth.

8.10.9 Oxygen

Microorganisms can be classified on their ability to use free oxygen: aerobes, anaerobes, facultative anaerobes.

Aerobes (aerobic) require free oxygen such as is found in the air, and without this supply they cannot grow.

Anaerobes (anaerobic) grow best in the absence of free oxygen. They satisfy their need for oxygen chemically through organic or inorganic compounds in their environment.

Facultative anaerobes will grow well either with or without free oxygen. Their varied reactions can confuse the novice. Certain facultative anaerobes can cause one type of change on the surface of a product, but can be a source of an entirely different reaction on the interior of the same or different product. Facultative anaerobic bacteria are important in changes produced in meat. The sausage greening organism ("green rings") occur in freshly cut sausage–the center core is not green due to lack of oxygen, but turns green after the bacteria (which have been growing anaerobically) are exposed to air and start aerobic growth.

8.10.10 pH (Hydrogen-ion Concentration)

pH is used to demonstrate the intensity of acidity or alkalinity. Every microorganism has a minimal, a maximal, and an optimal pH for growth. Foods with low pH values (below 4.5) usually are not readily spoiled by bacteria. A food, therefore, with an inherently low pH would tend to be more stable microbiologically than a neutral food. The same is true for protein meals and accounts for the desire of many renderers to have the Food and Drug Administration (FDA) in the United States of America approve of the use of acceptable organic acids for the industry. It is also desirable to know how some acids influence pH changes, because some acids, particularly the organics are more inhibitory than others.

Thus, not only are the rates of growth of microorganisms affected by pH, but so are the rates of survival during storage, heating, drying, and other forms of processing.

8.10.11 Moisture

Microorganisms have an absolute demand for water, for without water, no growth can occur. This is one of the classic absolutes of microbiology and underlies the insistence that moisture be controlled in, around, and on products to prevent bacterial replication and growth. The air, moisture, and relative humidity can also affect bacterial growth. Removal of moisture from floors, walls, and other surfaces can make a significant difference in bacterial buildup.

8.10.12 Temperature

The temperature at which microorganisms live or die is without doubt the most important single factor that determines their fate. Since microorganisms differ so widely in their minimal, maximal, and optimal temperatures for growth, it is obvious that the temperature will greatly influence the kind, rate, and amount of microbially induced change in rendered protein meals and cooked food like sausages. It is noteworthy that even a small change in temperature could favor an entirely different type of bacterium and result in a different type of spoilage.

Based on temperature characteristics, bacteria can be classified into 3 main groups:

1. Psychrophiles–"cold-loving"–can survive at temperatures below 20°C (68°F) and many grow well at refrigeration temperatures.

2. Mesophiles prefer warmer temperatures –21 to 43°C (70 to 110°F); body temperature is ideal for their growth. This encompasses the largest group of bacteria and includes the foodborne and other disease-causing bacteria.

3. Thermophiles are organisms that prefer temperatures from 40 to 60°C (120 to 140°F) or higher.

It is important to consider time and temperature together. These two factors are very important in the food, feed, and rendering industries. For example, pasteurization of milk can be done in seconds or minutes totally dependent on the time and temperature used. In the rendering industry, the time-temperature range will inactivate disease-causing bacteria, molds, yeasts, and viruses.

8.10.13 Yeasts

The term yeast, like mold, is commonly used but hard to define. Yeasts are classified chiefly on their morphological characteristics, although their physiological ones are more important to the food, feed, and rendering industries. Yeasts, exist as cells, as do all microorganisms. They are facultative anaerobes capable of growing with or without air. They multiply by a process called budding in which a small lump or bud appears on the side of the cell. This small bud grows until it is as large as the parent cell when it separates and becomes a yeast cell in its own right. Yeasts need moisture, warmth, and nutrition to grow and are exceedingly difficult to control, but they are not serious problems to the cooked food and rendering industries. They can grow at low temperatures and at acid pH. Products like bread and beer would not exist without yeasts.

8.10.14 Molds

The term mold is commonly applied to multicellular, filamentous fungi whose growth on foods is readily recognized by its cottony or fuzzy appearance. The main part of the growth usually appears white but may be colored or dark or smoky. Colored spores are typical of mature mold plants of some kinds and give color to part or all of the growth. Special molds are useful in the manufacture of certain foods or ingredients of foods. This is best exemplified by cheeses that are mold-ripened like Roquefort or Camembert; in Oriental foods like soy sauce and miso in breadmaking using amylase; and citric acid in soft drinks. Some molds produce toxic metabolites like mycotoxins which are highly toxic to many animals and potentially toxic to human beings.

Molds need air for growth and they grow on surfaces only. They can grow at low temperatures even below freezing and can grow well in acid pH. Molds will grow in low moisture and in the presence of high levels of salt. Molds are generally not considered harmful and they cannot compete with the common spoilage bacteria. Molds are very susceptible to high temperatures and can be easily killed by hot water.

Molds are easily carried by wind currents and moisture, and a low level of mold contamination is difficult to prevent in food, meat and rendering plants. In general, however, mold growth is not a problem in these industries.

8.10.15 Viruses

Viruses in general possess the following common characteristics:

- They are ultramicroscopic in size, varying from 10 to 450 nm.

- They can pass through most bacterial filters.

- They are cultivatable only on a susceptible host cell line.

- They are incapable of reproduction without a host.

- They can infect people, animals, plants, or bacteria; however, they tend to have a very specific target host.

The significance to the food, meat and rendering industries, including the world of medicine in general, is that we have only started to learn a lot more of the ecology and behavior of viruses. The viruses that cause disease in animals appear to be readily inactivated or killed by high temperatures like in the rendering process, and research projects clearly indicated that this happens with the virus that causes pseudorabies in swine. We think other viruses would probably succumb to the time-temperature regimen of the rendering process.

References

1. Mann, I (1984) : Guidelines on small slaughterhouses and meat hygiene for developing countries (VPH/83.53). World Health Organisation, Geneva

2. Goodman, J (1991) : Private communication. Meat and Livestock Commission, Milton, Keynes, U.K.

3. Campbell, S.C (1995) : Private communication. College of Veterinary Medicine, Cornell University, Ithaca, New York, 14353, USA.

4. Otto H. Siegmund et al. (1982) : A Handbook of Diagnosis and Therapy for the Veterinarian. The Merck Veterinary Manual.

5. Franco, D.A (1997) : Sanitation and Hygiene in the Production of Rendered Animal Byproducts under the auspices of the Animal Protection Producers Industry. The Fat and Proteins Research Foundation and National Renders Association.

CHAPTER 9

BIOGAS FROM ANIMAL WASTES

9.1 General

In nature "biogas" "natural gas" or "marsh gas" is a product of decomposition of organic matter such as leaves, grasses, wood, twigs and animal tissues buried under marshy places and anaerobic conditions. It is also found as a product of decomposition of the feed material in the stomach of the ruminants. Biogas is mainly composed of methane and carbon dioxide with small percentages of other gases. Methane, in practically pure form, occurs in coal mines and petroleum wells. At ambient temperature and pressure, methane is a colourless and odourless gas and is about half as heavy as air. In admixture with air, both methane and biogas burn with non-luminous flame and so the interest in commercial production of biogas from biomass or the organic matter. In biogas, methane is the only fraction which acts as a fuel; other gases being almost inert.

Anaerobic (i.e., without oxygen) treatment is based on a very simple tenet of life – every living thing must consume matter for existance. In this case, organisms utilize the carbon, commonly known as biochemical oxygen demand (BOD) in the waste water. These organisms have two sets of bacteria, being

(i) acid forming or liquifying

(ii) gasifying or methane producing

The first set (i.e., acid forming) converts carbohydrates, proteins and fats into volatile acids and in the process produces carbon dioxide as well. Liquefaction is brought about by a set of saprophytic bacteria by means of extra cellular enzymes. Being not very sensitive, these bacteria thrive over a wide range of circumstances. After the liquefaction is over, the methane producing bacteria take over converting the acids formed into methane and carbon dioxide with the help of intracellur enzyme. These bacteria are very sensitive to temperature and pH which are to be maintained at optimum for best results.[1]

Methane is the first member of a large family of chemical compounds called saturated hydro-carbons or paraffins. They are made up of two elements–carbon and hydrogen. Methane contains one carbon atom and four hydrogen atoms and is represented by the chemical formula CH_4.

9.2 Animal Wastes - An Ideal Source for Biogas Production

The typical raw materials as available from a slaughterhouse are organic in nature consisting of dung, urine, rumen contents, intestinal contents, blood, meat trimmings etc. Dung, rumen contents, and intestinal contents are not only rich in carbohydrates but also have fairly good amount of proteins. Others like blood, meat trimmings and urine are an excellent source of protein and nitrogen. This apart, the dung, ruminal contents and intestinal contents are also rich in micro-flora which help in the decomposition of these materials. Moreover, the slaughterhouse is a continuous or renewable source of supply of these raw materials.

Under chapter 12 on effluents, it has been discussed at length that proteins, fats and carbohydrates are easily converted into methane under anaerobic conditions. It will, therefore, obtain that slaughterhouse wastes are ideally suited for methane production; the chemistry of the bio-degradation of the organic matter under anaerobic conditions has also been discussed and does not need to be repeated. One is advised to refer to chapter 12 for more information on the mechanism of decomposition of different components.

9.3 Yield of Biogas from Different Types of Raw Materials

The yield of biogas depends upon the nature of raw material i.e. poultry, piggery, cattle manure etc. This is well illustrated by the approximate quantity of biogas as available from 1 kg. of manure each from poultry, pig and cattle, shown in Table 9.1.[2]

Table 9.1 : Yield of biogas from poultry, pig and cattle manures

Source	Bio-gas yield
1. Poultry manure	0.254 - 0.3 m³ (8.8 - 11.00 cft.)
2. Pig manure	0.045 - 0.064 m³ (1.55 - 2.21 cft.)
3. Cattle manure	0.026 - 0.037 m³ (0.9 - 1.28 cft.)

This difference in the production of biogas can be easily explained. The poultry feed contains very little fibre and more of proteins and starches. As such, manure obtained from poultry has more of easily bio-degradable components. Cattle feed, on the other hand, has considerable quantity of cellulose – a material which is not easily bio-degradable. Pig manure falls in between the poultry and cattle manure in terms of bio-degradable components.

9.4 Composition of Biogas

The biogas as produced, has a varied composition. There are two main gases of decomposition, which being methane and carbon dioxide with a few other gases in small percentages or even traces. Depending upon the raw meterial, temperature, pH etc. under which the reaction has been carried out, the percentage of methane in the biogas produced varies between 53 and 70 per cent.

9.5 Optimum Parameters for Anaerobic Fermentation

For best results, the concentration of the solids in the slurry to be fermented is around 10 per cent, temperature between 25° C – 35° C (the process of digestion and gasification being highest at around 35°C and almost coming to a stand still below 15°C) and pH around 7.5. It is also generally accepted that for optimum results, the ratio of carbon to nitrogen present in the bio-mass to be fermented should be around 30:1.[1,2] The raw material used should be well dispersed in water and the slurry in the digester is agitated periodically. This helps in breaking of the scum which forms at the top of the slurry in the digester.

Commercial systems are now available[3,4] where various parameters are maintained within reasonable limits. The temperature, in cold climates in particular, is regulated by utilising part of the methane produced, for heating of the slurry which is maintained at around 35° C for best results. The slurry is also agitated by bubbling methane through it from the bottom of the digester.[3] It may be worth mention that waste water from abattoirs, meat processing plants and rendering operations is comparatively warm because of use of steam and hot operations at some stage or the other. Hence, not much of energy may be needed to maintain desirable temperatures in the fermentation tank even in cooler climates.

9.6 Phases of Biogas Production

There are two phases of biogas production – the rapid decomposition of the more vulnerable components, chiefly the carbohydrates, followed by a comparatively slower production of gas from the less fermentable materials. Depending upon several factors, the rapid phase lasts anywhere between 15-25 days, followed by the slower phase extending even beyond 60 days.[1,2,5] In actual practice, the raw material being fermented may not be retained for all the two phases to be completed; rather the process may be terminated at some point during the slower phase of the cycle.

9.7 Biogas Plant

The biogas plant consists mainly of three parts – a digester or a set of these, where organic matter is digested; the gas holder, or a set of these, where the biogas is collected and stored; and the piping system which helps in the transportation of the gas to the place of its use. Depending upon the manner in which the gas is to be used, there may also be auxiliary scrubbers for carbon dioxide and hydrogen sulphide present in the gas produced.

The digester is a tank, normally either rectangular or circular in shape. It may be constructed underground or overground. Underground construction provides some insulation as well as ensures protection against sudden changes in atmospheric temperature. It should be airtight not only for the escape of the biogas it produces but also to prevent the entry of air into it, since the methane producing bacteria thrive only in the absence of oxygen. The digester should also be water proof.

There are two types of gas holders – fixed dome type and floating dome type. In fixed dome type, the biogas accumulates in the dome over the digester slurry (Fig. 9.1) The slurry serves as a reversible displacement medium. The biogas accumulating in the dome pushes out a portion of the slurry into a higher auxiliary compartment. The digester slurry flows back into the dome by gravity as the biogas is consumed.

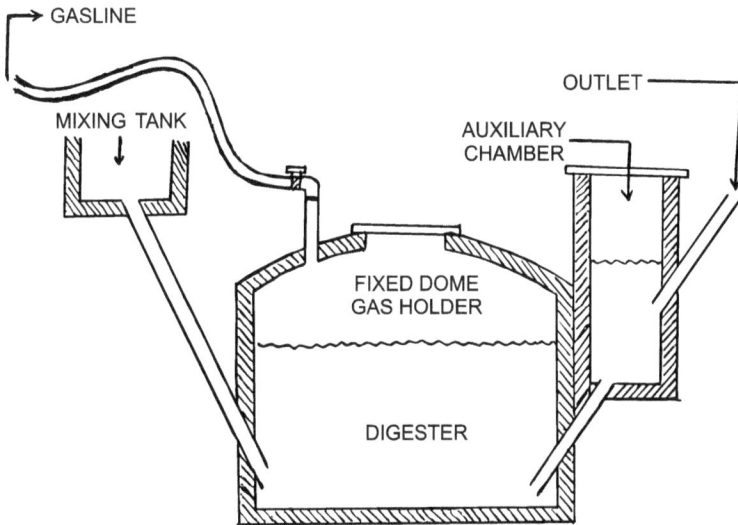

Fig. 9.1 : Chinese type of a biogas plant with a fixed dome gas holder.

The floating dome type consists of two parts; an open receptacle filled with liquid and an inverted tank floating on the liquid (Fig. 9.2).

There are two types of biogas plants – continuous type and batch type. In the continuous type, the raw material is fed daily to the plant and production of gas and the fermented slurry is continuous. The newly charged material automatically expels an equal volume of fermented slurry from the plant. In the batch type, there may be several digesters, each producing gas for a certain period after which it is partially or totally emptied and refilled with new gas producing material.

A number of biogas plant designs such as Indian, Chinese, Taiwanese etc. are available.[1, 2, 3, 4, 5, 6, 7, 8, 9, 10] Every design has its own merits and demerits and the choice of a particular design will depend upon the nature of raw materials used and other local conditions. For want of space, it may not be possible to discuss details of all the available designs. However, the design details of one of the most popular continuous type of Indian designs will be discussed.[7]

Cross section of a 6 m³ capacity continuous gas producing plant is shown in Fig. 9.2.

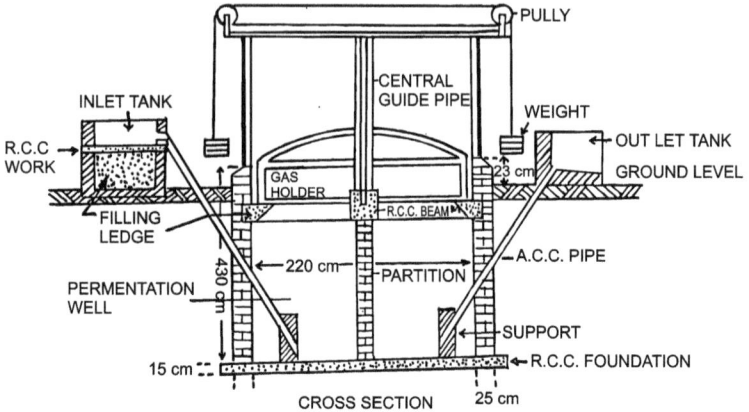

Fig. 9.2 : Cross section of a 6 m³ Indian continuous type biogas plant.

Capacity is defined to be the gas produced in cubic meters/ day at a temperature varying between 25–35° C. The design is based on cow dung to be used as the raw material. In other words, higher quantity of gas will be produced/day in the same plant if hog manure is used in place of cow dung.

9.7.1 The Digester

The digester is a well like construction. The wall of the well is made with brick and mortar, with the thickness of the wall being around 23 cm. Wherever available, stone blocks could be used in place of bricks. Stone is a better material against the seepage of water. The well is constructed under the ground and is provided with reinforced concrete and cement (RCC) foundation of a minimum thickness of 15 cm. Being under the ground, the well (digester) gets sealed against variations in atmospheric temperatures. The inner diameter of the well is around 2.3 meters and the total depth being about 4.3 meters. The well is divided into two semi-circular compartments by a partition wall constructed in the middle of the same. The partition wall prevents the mixing of the fresh slurry with the old fermented one. This is desirable for better efficiency of the plant. A ledge at a depth of about 95 cm from the top end of the digester is constructed as an integrated part of the wall of the well.

9.7.2 Gas Holder

The gas holder is a cylindrical drum made out either of galvanised iron sheet or mild steel sheet welded into a drum (with one end open). The outer diameter and flange height of the drum are around 2 meters and 1.1 meters respectively. The drum with its open end as the bottom, is suspendd into the well where it dips into the slurry and rests on the ledge. As the gas is formed and fills the empty suspended drum, its upward and downward movements are facilitated by a guide pipe welded at the centre of the base of the drum. In fact, the welded pipe is sleeved on to another rod or pipe which has been embedded into a block of RCC constructed over the partition wall. A stone block can also be used in place of RCC block. The drum has been provided with a counterweight arrangement.

The gas produced in the well gets collected inside the drum. The pressure on the gas is equivalent to the weight of the drum and can be further increased by loading extra weights as desired, in the counter weight system. The pressure exerted on the gas helps in its transportation to the place of its use through the system of valves and pipes as shown in Fig. 9.2.

9.7.3 Inlet and Outlet Tanks

The inlet and outlet tanks are square shaped with wall lengths of 75 cm and 50 cm respectively. The height of the walls for both the tanks could be 50 cm in each case. The inlet tank is used to feed the slurry in the digester while the outlet tank receives the fermented slurry.

When cow dung alone is used, the same is converted into a slurry by mixing it with water approximately in the ratio of 1:1 before it is fed into the digester. In the case of other slaughterhouse wastes like urine, ruminal contents, intestinal contents, blood, meat trimmings etc., the percentage of moisture present is comparatively much higher than the dung alone. As such, while making slurry out of these materials, the quantity of water to be mixed is to be adjusted. The guiding factor is that percentage of solids in the slurry is adjusted to around 10 per cent both in the case of dung or other materials.[4, 7] Care should be taken that no solids (like big and long pieces of straw from the lairs or rumen contents) are added to the slurry. These settle down at the bottom and reduce the usable space in the digester. The material should be well dispersed in the slurry; finer the size of the material, better the results. The slurry is fed through the inlet tank. A wire gauge strainer should be used to retain any big size material present in the slurry, from entering into the digester. As the digester is fed with the fresh slurry, an equivalent volume of the old fermented slurry is expelled out through the outlet pipe.

9.8 Products of Biogas Plant

There are two main products available from the biogas plant - the fermented slurry and the biogas.

9.8.1 Biogas

Depending upon the nature of bio-material used, the temperature, pH and other conditions of fermentation, the composition of the biogas produced may vary as follows:[1,2,6,8,9]

Methane	53 - 70 per cent
Carbon dioxide	27 - 43 per cent
Hydrogen	1 - 10 per cent
Nitrogen	1 - 5 per cent
Oxygen	0.5 - 1 per cent
Carbon monoxide	0.1 per cent
Hydrogen sulphide	traces
Others	traces

The most efficient use of biogas is in cooking or for direct heating purposes. With properly designed burner, as much as 60 per cent fuel value of the gas could be utilized for heating.

Biogas is nonpoisonous with a characteristic odour which disappears on burning. In admixture with air, it burns with a clear, bluish and soot free flame. In its original form, it has a very low inflammability and a burning cigarette will not ignite it. However, it is flammable in admixture with air at a level of around 8-20 per cent of the biogas in the mixture.[1,2,6,8,9]

Pure methane burns in a mixture of 91 per cent air and 9 per cent methane. The biogas, however, needs approximately 93 per cent air for it to burn completely. For this reason, the normal LPG burners used commonly in the houses are not fit for biogas. Special burners with bigger holes have been designed for use of biogas as a fuel. Bigger holes enable to draw higher percentage of air from the atmosphere, needed for efficient burning of the biogas.[11] With the correct air and biogas mixture, the flame temperature can reach as high as 800° C.

To get an appreciation of the quantity of raw materials required and the volume of gas needed for various end uses, some figures are provided. As already discussed, one kilogram of cattle dung produces around 0.037 m^3 (1.3 cft.) of biogas at atmospheric pressure and roughly 25° C temperature. Corresponding figures for poultry and pig manures have also been given (Table 9.1). The availability of dung from an average sized stall fed cattle is around 10 kg/day. Hence, dung obtained from at least 5 cattle heads will produce gas equivalent to around 1.7 m^3 (60 cft.). Under Indian conditions, this volume of gas is considered to meet the cooking and lighting needs of a family of 4 persons per day.[4]

For using the biogas for lighting purposes, special mantle lamps have to be used. A 100 candle power lamp, roughly consumes 0.13 m^3 (4.5 cft.) of the gas/hour.[4]

As for power, internal combustion engines could be converted for use of biogas as fuel. For doing so, a special attachment is required to be attached to the combustion engine. Before use, the traces of hydrogen sulphide (H_2S) should be removed from the biogas; if not, it will corrode the engine. For this, the biogas is passed through ferric oxide or iron filings which help in scrubbing of the hydrogen sulphide.[2] When ferric oxide is used, the same can be regenerated by heating it in the presence of air and reused. About 0.48 m^3 (17 cft.) of fuel gas is required to run an engine of one horse power for one hour.[2,4]

For gasoline engines, biogas can completely replace gassoline. Gasoline engines could be converted for use of biogas by changing the carburetors.

In the case of diesel engines however, diesel can not be replaced totally with biogas. Some amount of diesel is required for ignition. The diesel engine is converted to a dual fuel (diesel + fuel gas) engine by attaching a biogas carburetor to the air-intake manifold. When biogas is used, the consumption of diesel can be brought down to about 15-25 per cent.

Not only that biomethane can beautifully replace conventional fuels, it is a giant green step forward. According to one unpublished report, if used as a replacement of diesel, biomethane is said to emit 78 per cent less nitrogen oxide and 98 per cent fewer fine particles– two causes of respiratory illnesses and 92 per cent less noisy.

9.8.2 The Slurry

During the past, slurry was considered as a product only fit to be used as manure either fresh in liquid form diluted with water or after drying as powder.

However, the excellent work which has been carried out at Maya Farms, in Phillipines, has opened a number of other important avenues for its use.[2]

The slurry is an aqueous suspension of solids in a liquid. It resembles dark mud in appearance and contains large quantity of fine fibres.

When fresh, it still produces some biogas. However, as the slurry is exposed to atmosphere, the methane formation stops; the same can take place only under anaerobic conditions. The fresh slurry emits a smell of hydrogen sulphide and is not any more too offensive as is the case with fresh dung.

As the formation of methane stops, the slurry separates into two portions – the dark solid portion settles at the bottom leaving a turbid liquid sludge layer at the top. The liquid portion of the slurry, also called liquid sludge is run out into another lagoon leaving the solid sludge behind.

9.8.3 Solid Sludge*

The solid sludge which has settled at the bottom is scooped out and dried in open sun. When analysed, the solid sludge is found to be rich in protein (around 20%), fat (around 5%), fibre (around 10%), phosphorous, potassium, calcium, magnesium, iron,

---

* The data given is in respect of sludge obtained from pig manure.

copper, zinc and manganese. The sludge is also free from parasites. Preliminary tests also indicated it to be free of salmonella. It is also rich in vitamin B12. The analysis shows that the product is not only fit to be used as fertilizer but could better be utilized as a feed material in livestock feeds. Infact, the feeding trial tests on pigs showed a very good weight gain when the material was used at 10 per cent level of the total feed. This growth cannot be explained due to the protein alone. It is expected that vitamin B12 and perhaps other unidentified growth factors are responsible for the growth.

9.8.4 Liquid Sludge

About 90 per cent of the total slurry comprises of liquid sludge. As such it poses a big disposal problem when the size of plant is large.

The liquid portion, when analysed was also found to contain nitrogen, phosphorous, potassium, calcium, magnesium, iron, copper, zinc, maganese, etc.

At the Maya Farms, the liquid sludge has been successfully used for the following.

1. *As irrigation water and fertilizer:* When used for irrigating paddy fields, blue-green algae have been reported to grow in profusion which in turn help in fixing atmospheric nitrogen. The plants irrigated with the liquid sludge were found extremely healthy and it is suspected that other than supplying nitrogen and minerals to the plants, the sludge may also be containing plant hormones and unidentified growth factors.

2. *For fish culture:* When exposed to the sun light, the sludge is reported to support profuse growth of plankton. Some of the fish, tilapia in particular, feed on this plankton. As such, liquid sludge has been found very useful for fish culture.

3. *For algae culture:* The liquid sludge has also been found suitable for growing Chlorella and other similar algae. However as algae culture also involves further additional operations for its harvesting, drying and bagging, economy of algae culture remains to be established.

9.9 Toxicity of the Slurry

If used for irrigation and fertilizing purposes, the literature on biogas states that the slurry should be allowed to "ripen" before use. When used fresh, the slurry is toxic both to fish and plants. It has been observed at Maya Farms that even very hardy varieties of fish and plants do not survive in the fresh slurry even when highly diluted. It is believed that the toxicity is primarily due to hydrogen sulphide, even if present in traces in the slurry.[2]

When aerated, the hydrogen sulphide gets oxidised to water and elemental sulphur. Whatever may be the reason of toxicity, it has been found that well aerated liquid sludge is no longer toxic to fish and plants. Similarly, solid sludge becomes safe for use when the same is exposed to air and dried.

It is therefore important that the liquid slurry is first aerated well in lagoons before the same is used either for irrigation, production of algae or fish culture; aeration may be hastened if the surface area of the lagoon is increased or by bubbling in air with the help of wind mill etc.

9.10 Composting

Animal wastes like ruminal contents, dung etc. are many a times composted as discussed in chapter 10 and the compost used as manure. It may therefore be worth discussing the merits and demerits of making use of the animal wastes for compost making.

Considerable quantity of nitrogen is lost in the form of ammonia during the process of composting. This apart, soluble nitrogen such as in urine, even gets leached out during the process of composting.

As against this, when the animal wastes are fermented under anaerobic conditions as discussed, the loss of nitrogen is practically negligible. The solid portion of the fermented slurry is rich in proteins, vitamins and minerals. It is also free from salmonella and has been found to be extremely useful as a protein feed supplement in livestock feeds. Similarly, the liquid portion is rich in nitrogen and minerals and is useful for irrigation, fish culture and growing of algae. Wherever possible, the animal wastes should therefore be utilized for biogas production. Composting of these wastes is utter under utilization.

References

1. Srinivasan, H.R (2001) : Energy from carcass wastes, in *"Technologies for value realization of carcass byproducts in developing countries – a handbook"*. Central Leather Research Institute, Chennai-400020 (India).

2. Felix D.Maramba, Sr. : *Biogas and waste recycling – the Philippine experience*. Maya Farms Division, Liberty Flour Mills, Inc. Metro Manila, Philippines.

3. Farm Gas Ltd, Industrial Estate, Bishop's Castle, Shropshire Sy9 5AQ, UK.

4. *Gobar gas, why and how.* (1994) : Directorate of Gobar Gas Scheme, Khadi and Village Industries Commission, Gramodaya, Irla Road, Vile Parle (West), Bombay.

5. Ritsema, M. (2007) : Biogas. Render, June, 2007.

6. Mahendra Kumar (2007) : *Animal byproducts utilization through semi-moist rendering*. Daya Publishing House, Delhi-110035.

7. IS : 9478 – 1980 : *Specification for gobar gas plant and equipment (with amendment No. 1, January 1983).*

8. Mahendra Kumar (1989) : *Handbook of rural technology for processing of animal byproducts (FAO Agricultural Services Bulletin No.79).* Food and Agriculture Organisation of the United Nations, Rome.

9. Mahendra Kumar (1987) : *Processing of animal byproducts in developing countries – a manual.* Commonwealth Science Council, Marlborough House, Pall Mall, London SW1Y 5HX, UK.

10. Mann, I (1962) : *Processing and utilization of animal byproducts (FAO Agricultural Development Paper No. 75).* Food and Agriculture Organisation of the United Nations, Rome.

11. IS : 8749 – 1978 : *Specification for gobar gas stove.*

CHAPTER 10

COMPOSTING AND VERMICOMPOSTING

Composting including vermicomposting of biodegradable segment of substrate are of utmost importance to keep the environment clean. This apart, these options are also a source of some income and so also solution to a very critical problem of finding a local solution for the production of quality manure for sustainable agriculture.

Both theoretical as well as practical aspects of making compost and vermicompost will be discussed.

10.1 Composting-Desirable Parameters

Essentially, composting is a decomposition process of organic material; decomposition can be carried out both under aerobic and anaerobic conditions but for composting, decomposition under aerobic conditions is a preferred one. All kinds of organic wastes, both of plant and animal origin, can be composted. The process is carried out in solid state system by an aerobic community of microbes which also include mesophilic and thermophilic populace. In other words, the composting process seeks to harness the natural forces of decomposition.

For best results, certain parameters have to be at their optimum which being:

1. porosity and aeration
2. moisture content
3. carbon, nitrogen and carbohydrate ratio
4. pH
5. atmospheric temperature

Each of these parameters are explained.

10.1.1 Porosity and Aeration

Porosity and aeration are inter-related. The process of composting being aerobic, good amount of air is needed for it to proceed successfully; porosity provides just that. It facilitates circulation of air in the compositing pile.

The aeration serves more than one purpose – it provides oxygen for continued microbial respiration and removes the end products of microbial metabolism *viz.* carbon dioxide and water. If the latter accumulates, it can make the mass too wet and if this happens, or if the water content of the mass is initially too high, undesirable anaerobic microbial populations becomes dominant.

10.1.2 Moisture

The moisture content of the pile has to be around 50–60 per cent; if it is too high, undesirable anaerobic microbial pupulace becomes dominant leading to anaerobic decomposition and formation of methane. In case the moisture is low, the aerobes do not get enough moisture for the activity.

10.1.3 Carbon, Nitrogen and Carbohydrates

For any microbial activity, carbon, carbohydrates (source of energy) and nitrogen (*e.g.* protein and ammonia) are essential; ideally these have to be in such proportion that carbohydrates are just enough as needed for the conversion of nitrogen present in the waste into microbial nitrogen. Carbon (C) to nitrogen (N) ratio of around C:N = 26–30:1 is considered to be most desirable. Since the animal wastes are of very complex nature, such ideal conditions are rare to be achieved. All the same, reasonably good results are possible to achieve by following a set of practices worked out over the time. These will be described later.

10.1.4 pH

The pH of the composting mass should ideally be neutral (*e.g.* 7) but may vary between 6.5 to 7.6. Very acidic or alkaline systems are not desirable for composting.

10.1.5 Atmospheric Temperature

External or atmospheric temperature is of interest in as much as it will have significant influence on the composting process.

Warm climates are preferable for composting than the colder ones. Since composting process passes through mesophilic and thermophilic cycles, cold atmosphere temperatures will naturally delay the composting process.

10.2 Composting–Basic Dynamics

In a properly composted system, temperatures around 70°C are achieved. In the first phase, mesophilic organisms are active which heat up the mass upto around 44°C and consume most readily available carbohydrates; the temperature continues to increase and in the process they get increasingly inhibited. In the transition range from about 44°C to 52°C, the thermophilic organisms begin to grow and continue to heat the mass. Thermophilic growth then continues to 75°C. Most of the pathogenic and thermolabile microorganisms are destroyed or reduced during this period. Other contaminants like weeds seeds, organic and organometallic compounds also get degraded.

After the thermophilic cycle, the system starts cooling. New bacterial activity starts as soon as an appropriate temperature is reached. Turning of the compost at this stage not only cools the system but also helps in triggering microbial activity and growth, resulting in the use of readily available nutrients and increase in temperature. As the nutrients are progressively used and microbial activity declines, the compost again cools down. Thereafter, microbial activity again starts and more complex substrates present in the mass, such as cellulose, lignin etc. are utilized; these substrates being complex in nature are acted upon at very slow rate and the heat generated in this activity is lower than the heat loss in the atmosphere. Hence the compost continues to cool down. The active organisms in the cooling phase are usually different from those involved in the heating up phase; the nature of nutrients as available during the heating up and cooling down phases is also found to be different. The final stage in compost making is that of maturing; in this phase macro-organisms such as worms and insects appear in the system and there is considerable predation between the various biotic components of the compost. Typical macrofauna which appears at the maturing stage may consist of mites, ants, termites, millipeds, springtails, centipedes, spiders, beetles and worms like earthworms. Macrofauna plays a very vital function in the composting process; it breaks the substrate into smaller pieces resulting in increased surface area. This facilitates the microorganisms to bring about further breakdown of the matter through excretion of some chemicals and enzymes. This aparts, the excreta of the microfauna enriches the quality of the compost; attributes of vermicast have been discussed under vermicasting.

In practice the composted material is considered mature when it fails to heat up on turning and does not go anoxic.[1] Time and temperatures required for the destruction of some of the commonly encountered pathogens and parasites in a compost-stack/heap are listed in Table 10.1.[2]

Table 10.1: Temperature and time of exposure required for destruction of some common pathogens and parasites in a compost stack/heap

Organism	Observations
Salmonella typhosa	No growth beyond 46°C, death within 30 minutes at 55°C–60°C and within 20 minutes at 60°C
Salmonella sp.	Death within 1 hour at 55°C and within 15–20 minutes at 60°C
Shigella sp.	Death within 1 hour at 55°C
Escherichia coli	Most die within 1 hour at 55°C and within 15–20 minutes at 60°C
Entamoeba hytolytica cysts	Death within a few minutes at 45°C and within few seconds at 55°C
Taenia saginta	Death within few minutes at 55°C
Trichinella spiralis larvae	Quickly killed at 55°C and instantly killed at 60°C
Brucella abortus or Br. Suis	Death within 3 minutes at 62°C–63°C and within 1 hour at 55°C
Micrococus pyrogens var. aureus	Death within 10 minutes at 50°C
Steptococcus pyrogenes	Death within 10 minutes at 54°C
Mycobacterium tuberculosis var. homins	Death within 15–20 minutes at 66°C or momentary exposure at 67°C
Corynebacterium diptheriae	Death within 45 minutes at 55°C
Necator amercarnus	Death within 50 minutes at 45°C
Ascaris lumbricoides eggs	Death in less than 1 hour at temperatures over 50°C

As numerous kinds of microbes are involved in the process of composting, it may be desirable to briefly discuss the individual role of all the microbes involved in the system *viz.*, bacteria, fungi, actinomycetes and protozoa.

10.3 Microbes

Bacteria, fungi and the actinomycetes are the main degrading microorganisms in a composting system; protozoa also have some role to play.

10.3.1 Bacteria

Bacteria are smallest (in size) and most numerous of the organisms involved in decomposition. They are single cell identities approximately measuring 1 micron in diameter and upto 10 micron in length; one micron being equal to 1/1,000 mm. They are differentiated by their shapes, being either spherical, rod shaped or long spiral; some of these can survive under unfavourable conditions by forming spores, which can withstand dehydration and high temperatures for long periods.

Under aerobic conditions, certain bacteria utilize atmospheric oxygen, decompose organic matter and assimilate some of the carbon, nitrogen, phosphorous, sulphur and other nutrients for synthesis of their cell protoplasm. In the process, production of carbon dioxide, humic substances and release of available plant nutrients takes place.

Various workers have isolated and identified the species of bacteria generally encountered in aerobic composting environment–thermophilic stage in particular, *Bacillus* spp. constituted the majority of isolates from materials being composted in the upper 50°C and low 60°C range but above 65°C, almost monocultures of *B. stearothermophilus* were encountered. The results of these studies have been summarized and published;[2] same are reproduced below:

Species of Bacteria

Bacillus

   B. brevis

   B. circulans complex

   B. coagulans type A

   B. coagulans type B

    B. coagulans type C

    B. licheniformis

    B. sphaericus

    B. stearothermophilus

    B. subtilis

Clostridium

    C. thermocellus

    Clostridium sp.

Pseudomonas

    Pseudomonas sp.

Additional information on bacteria is available in Chapter 8.

10.3.2 Actinomycetes

Some bacteria have a tendency to form filaments. Such forms of bacteria are called actinomycetes. Actinomycetes share characteristics both of bacteria and fungi. They are in an evolutionary phase between the two and are next to bacteria in numbers. Like bacteria, they are found more commonly in neutral to slightly alkaline environments. They occur most commonly and abundantly inside freshwater, manure and substrate rich in dead organic matter. They are also able to withstand stressful conditions.

Like bacteria, various workers have isolated and identified *actinomycetes* in the aerobic composting piles; the list of thermophibic actinomycetes commonly recovered have been consolidated and published;[2] the summary is given below:

Species of Actinomycetes

    Actinobifida chromogena

    Microbispora bispora

    Micropolyspora faeni

    Noncardia sp.

    Pseudonocardia thermophilia

Streptomyces

    S. rectus

    S. thermofuscus

    S. thermoviolaceus

    S. thermoulgairs

    S. violaceus-ruber

Streptoymces sp.

Thermoactinomyces

    T. vulgaris

    T. sacchari

Thermomonospora

    T. curvatea

    T. viridis

Thermomonospora sp.

10.3.3 Fungi

Fungi are hetrotrophs and therefore derive nutrition from either living plant tissues as parasite or dead tissues as saprophytes. Being filamentous, fungi are able to penetrate into the rigid/hard tissue like lignin and produce hydrolytic enzymes causing local softening of tissues. Saprophytic fungi alongwith certain bacteria act as major decomposers of organic matter.

The consolidated results of various workers connected with thermophilic fungi as recovered in a composting environment have been published;[2] the same are reproduced below:

Species of Fungi

    Zygomycetes

    Absidia

    A. ramosa

    Absidia sp.

    Mortierella turficola

    Mucor

    M. miehei

    M. pusillus

    Rhizomucor sp.

    Ascomycetes

    Allescheria terrestris

Chaetomium thermophilum

Dactylomyces crustaceous

Myriococcum albomyces

Talaromyces (Penicillium)

T. dupontii

T. emersonnii

T. thermophilus

Thielavia

> *T. thermophila*

Basidiomycetes

Coprinus

C. lagopus

Coprinus sp.

Lenzites sp.

Deuteromycetes

Aspergillus

> *A. fumigatus*

Humicola

> *H. grisea*

> *H. insolens*

10.3.4 Protozoa

Protozoa are single cell organisms and are larger in size than most other organisms. The life cycle of protozoa has two phases of life cycle being:

(1) active phase (multiplication phase) and

(2) resting phase

The later takes place under adverse environmental conditions like low moisture, high temperature or both, wherein it forms a thick coating around itself to be called "cyst". Most of the protozoa in composting matter are saprophytes. Some of the protozoa directly feed upon bacteria present in the organic substrate. Their predatory nature keeps a check on bacterial population.

10.4 Materials for Composting

Practically all kinds of wastes as available at a slaughterhouse or rendering plant such as manure, urine, blood, ruminal and intestinal contents, meat and hide trimmings, floor sweepings, feather, hair and condemned meat could be used for composting. Condemned meat and trimmings etc. should, however be chopped and mixed with earth and evenly spread out at the centre of the compost heap, where the temperatures are the highest. Where quantities are small, chopping can be done manually. Size of chopped pieces of not more than 5 mm is desirable for proper composting.

Slaughterhouse wastes apart, some of the other organic roughages, coarse in nature, are essential for successful composting. These could be materials like maize and sorghum stacks, sugarcane trash, straw, wood shavings, broken baskets, waste paper, and the like.

Other materials like vegetable wastes, road sweepings, poultry and fish wastes, over ripe fruits, kitchen wastes and in fact, all kinds of organic materials could be suitably combined and converted into compost.

10.5 Methods of Composting

Composting is carried out by following any of the five methods listed below:

1. Pit Method
2. Stacking Method
3. Windrow Method
4. Vermicomposting Method
5. Mechanically Rotated Drum Method

Methods 1 to 4, are very simple, and no elaborate equipments are needed for the same. The first four methods will be discussed in detail. Fifth method is considered out of scope of this write-up.

10.5.1 Pit Method of Composting

Pit method is the simplest method of composting and is, invariably practiced to convert manure, urine and other materials into compost.

In this method, the materials to be composted are dumped into a shallow pit dug in the ground; complex microbial reactions take place until the material is composted.

Strictly speaking, this method cannot be called as true composting method, as no proper aeration of the mass takes place; it can at best be called as a rubbish dump.

All the same, microbial degradation of the material does take place, both aerobically and anaerobically under anaerobic conditions, some methane also gets generated. The system may still be useful where other methods could not be introduced, even though the end product does not compare favourably with a properly composted one but can still be used as manure. The practice, no doubt, helps in keeping the surroundings of slaughterhouse or rendering plant clean and tidy.

10.5.2 Stacking Method of Composting

This method consists of stacking or piling of vegetable matter and any offals in alternate layers above the ground. The process is best carried out in masonry bunkers; alternatively the stack could be covered with vegetable matter.

10.5.3 Construction of Bunkers

Stone blocks, cement blocks or bricks may be used for the construction of bunkers, raising the wall above the ground, leaving open spaces in the wall while the same is constructed. The open spaces help in air circulation through the pile.

The size of the bunkers will depend upon the quantity of the raw material to be composted. Figure 10.1 shows a layout which permits a lorry or a cart to reverse into the bunker for loading of the compost. Perspective view of the bunker is shown in Figure 10.2.

There are four bunkers (A, A1 and C, C1) for holding the raw material. Bunkers B and D are for holding partly composted material from bunkers A and A1 and C and C1 respectively; bunker E further holds composted material from bunkers B and D. The bold lines in the plan indicate masonry walls while white light lines in parallel indicate removable wooden partitions. The floor of all the bunkers is earth only. After the bunkers have been constructed, these are ready for use.

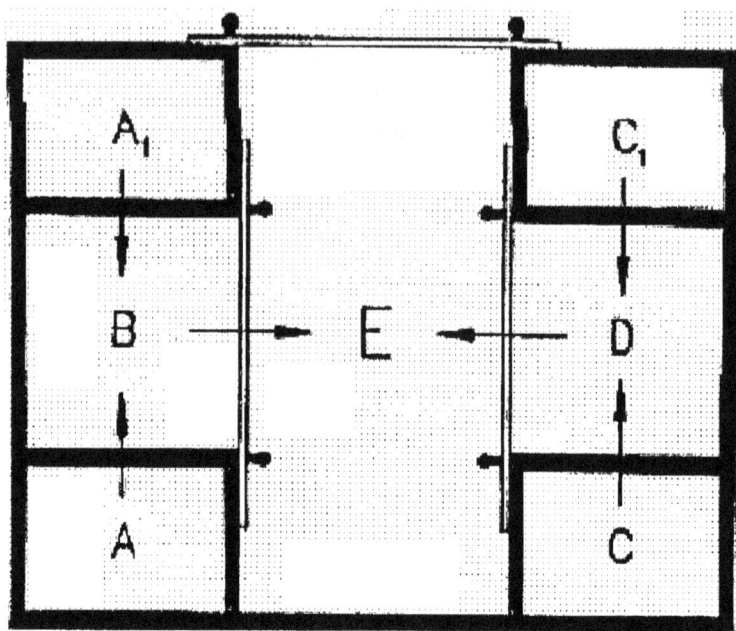

Fig. 10.1 : Plan of a Compost Bunker

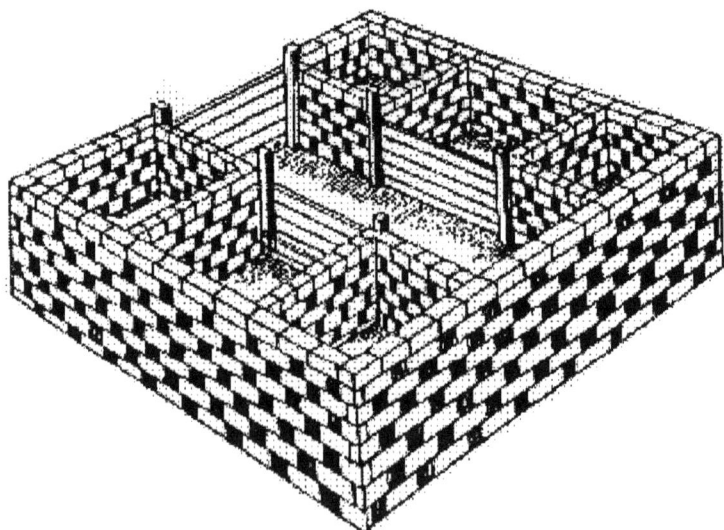

Fig. 10.2 : Perspective View of the Bunker

10.5.4 Loading

The raw material is loaded into bunkers A and A1 and C and C1. The first layer which forms the base of these 4 bunkers must be coarse vegetable roughage such as sorghum, maize or banana stalks. For good aeration, this layer should be about 15 cm thick, topped with a 8 cm layer of comparatively less coarse material like wheat straw, paddy straw, grass, leaves or wood shavings. This base layer is now loaded with materials like dung, ruminal and intestinal contents, offals or rejects if any, hair, feather etc. It will be desirable to mix all the materials before loading; the thickness of the layer being about 15 cm. This layer is now topped with vegetable waste matter. Such alternate layers of vegetable matter and materials from slaughterhouse and rendering plant are repeated until the bunker is full; the last layer being that of vegetable matter. The total height of the stack could be around 150 cm. To prevent drying, the bunker should now be covered suitably with materials like asbestos sheet or corrugated iron sheet – the former being preferable because of its non-rusting nature. When the material becomes too dry, the fermentation is slowed down and it may be necessary to water the stack now and then – in dry climates in particular.

The alternating layers of vegetable waste help in aeration and equalization of moisture as well as its retention all through the system.

10.5.5 First Turning

After a retention period of about 2.5 to 3 weeks, the contents of the bunkers A and A1, are turned and stacked together in bunker B. The contents of C and C1 are turned similarly and stacked in bunker D.

10.5.6 Second Turning

The stacks in bunkers B and D are retained for a further period of about 4 to 4.5 weeks, whereafter the contents of the two are turned and stacked together in bunker E; the mix is retained in bunker E for a further period of about 6.5 to 7 weeks. By this time (a total of about 90 days), the contents have decomposed sufficiently and are ready for use.[3]

It may be well appreciated as discussed earlier that composting is dependent upon a number of variables in the system *e.g.* nature, proportion of raw materials, water content, temperature, pH, porosity, air circulation etc. Hence the discussion as offered, could at best, be treated as a guide. One will learn from experience to suitably combine the available raw materials – both that of slaughterhouse as well as rendering plant and the vegetable matter, as to obtain the best possible results.

10.6 Stack Covered with Vegetable Matter

In this method, the stacking or piling of the available raw materials is done in the same fashion as explained for masonry bunkers; however, the heap is covered from all sides as well as on the top by a thick layer of vegetable matter. Coarse materials like tree branches, long grasses, maize and sorghum stalks and the like have to be used for covering. Clean earth could also be used for covering.[3]

The heap has to be turned at least thrice, as was the case in bunkers. While turning, thorough mixing of the contents should be assured to obtain a uniform product. If the process is carried out properly, the other details like the period after which the first and second turning are to be done, the total time taken for composting and so also the temperatures obtained in various phases of the process are almost the same as that in the case of bunkers.

10.7 Windrow Method

10.7.1 Introduction

Considerable inputs on the Windrow Method of composting are through the courtesy of Dr. S.S. Khanna who has indepth knowledge and expertise of the subject.[4]

The method can be considered as modified but improved method of stacking, known as "Windrow Method". The method has been widely followed for composting "Solid Urban Wastes". The main aim of the method is not only to improve the quality of the compost but also bring down the period of composting from around 90 days as discussed under 10.5.2 to about 6–7 weeks.

Period of composting is possible to be brought down by the following two interventions.

1. As also in the earlier process of composting, to keep the composting parameters like pH, moisture, temperature, porosity and aeration and C:N ratio of the substrate at the optimum level, these parameters have earlier been discussed in detail.

2. Addition of inocculant to material during composted.

Inocculant is a mix of microbes responsible for composting *e.g.* bacteria, actinomycetes, fungi and protozoa. The inocculant contains several genera and species of each of the four microbes. All these genera and species play their specific role during the composting process at one stage or the other and most of the species of microbes work in tendom with each other. The inocculant is available commercially in liquid form. One factor which needs consideration about the inocculant is its price which is considered to be high (in India), being around Rs. 50/litre (US$ 1/litre). Fortunately, however, the inocculant can be conveniently cultured in the laboratory, at home or even at the workplace itself. The process is very akin to the preparation of curd at home wherein a small quality of culture (curd) is added to reasonable large quantity of milk at the right temperature for it to be converted into curd.

10.7.2 Culturing of Inocculant

Fresh cow dung is diluted with water into slurry in a large drum. One litre of commercially available inocculant is added to the slurry and the contents mixed thoroughly as well as aerated. One very simple method of aerating is to take a mug full of the mix and pour it back from handful height into the drum itself; repeat this operation many times. This simple operation will add good quantity of air into the mix. As microbes are aerobic, their activity will accelerate when the air or oxygen is available in the mix. The activity can further be accelated with the help of a suitable catalyst; marigold flowers (available in yellow, maroon and orange colours) have been found to be good catalyst. The flowers are crushed and added to the aerated mix. The inocculant mass is ready for use within hours.

10.7.3 Method of Windrow Composting

The process is carried out on a cemented platform. The size normally being 20mt width and 30mt length. The platform has

1 per cent gradient from ends A and B towards ends C and D, followed with channel E. The channel is meant for collection of leachate from windrows. The leachate is an excellent inocculant and is made use of accordingly. The platform may be marked with packets of 2mt by 3mt in an arrangement shown in Figure 10.3

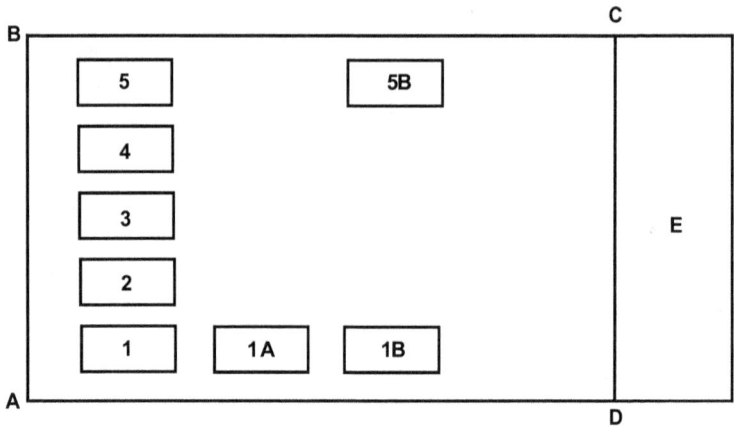

Fig. 10.3 : Arrangement of pockets on the platform and Channel E.

Pockets 1 to 5 is one set, similar pocket sets are in parallel to this set like 1A to 5A, 1B to 5B and so on. All pockests have not been shown in the figure. The material to be composted is stalked into a pile on pocket 1 measuring 2mt breadth, 3mt length and 2mt height. The pile is called "Windrow" and so the name "Windrow Method of Composting".

For best results, coarse materials like sorghum, maize, banana stalks etc. have to be a part of the softer materials like dung, rumen digesta etc. emating from slaughterhouses and rendering plants. The coarse material should be finely cut/shreded and thereafter this material and the soft one should be mixed thoroughly. The mix is now ready to be spread into a pile on pocket 1 in layers of about 15cm each. Each layer should be sprinkled liberally with the inocculant before spreading second layer on it until a height of 2mt.

The composting mass from pocket 1 is progressively transferred to pockets 2, 3, 4 and finally to 5; the mass remaining for specific periods in each pocket. The process of transfer from one pocket to another is known as "turning" in the composting parlance. The periods for which the composting process is carried out at each pocket are given below:

1. Pocket 1: Duration of holding, thereafter   7–10
             transferred to pocket 2   days

2. Pocket 2: Duration of holding, thereafter   17–20
             transferred to pocket 3   days

3. Pocket 3: Duration of holding, thereafter   7–10
             transferred to pocket 4   days

4. Pocket 4: Duration of holding, thereafter   7–10
             transferred to pocket 5   days

The process is considered complete by the time the mass is turned to pocket 5. Pocket 5 is more of a holding place, eventhough some decomposition may still be taking place.

The compost at this stage has around 50 per cent moisture and has to be spread in the open for the moisture to be brought down to around 15 per cent before packaging for sale.

Like earlier methods, the piles have to be suitably covered and sprayed now and then with water for the moisture to remain between 50–60 per cent.

Turning plays a very vital role both for the physical characteristics of the compost as well as biotic activity. It

(1) transfers the composting material upside down which helps it to get exposed to the meosphilic and thermophilic cycles all through.

(2) mixes the material as to get very uniform final product.

(3) breaks anaerobic pockets formed, if any, in the pile and

(4) aerates the material leading to good biotic activity.

Turning is best done with the help of a scoop attached to a tractor particularly if large quantities are handled. Manual turning is practiced for small lots but one is likely to get exposed to all types of microflora which may cause skin diseases. While doing manually, one should ensure the use of gum boots, gloves and masks.

It needs to be pointed out that in areas of high rain fall, the composting should better be done under sheds. If not, there will be considerable interference with biotic activity as well as leaching of the compost.

Particle size of the material to be composted has a profound influence on the speed of composting; smaller the pieces, faster it is. As most of the raw material from a slaughterhouse or a rendering plant is in the form of dung and rumen digesta, which have not only been well shreded as well as decomposed by the animal but also a large population of desirable microflora being present in it, the windrow method is expected to suit well for this raw material.

10.8 Benefits of Using Compost as Manure

It is a well-established fact, especially for tropical and subtropical soils, that composted manures improve the physical condition of the land. The product is a mixture of humus like particles useful as soil conditioner. Humus is a dark, biologically stable and finely divided amorphous organic matter which is decomposed plant and animal material rich in N, C, P and S content. Generally speaking, humic products are formed by the combination of aromatic compounds present in the organic matter, with the decomposition products of proteins and polysaccharides. Because of its high organic content, it helps to provide good tilth and water holding and nutrient retaining capacity when mixed with poor soil; as a result, soils otherwise unsuitable, can be put into cultivation by regular use of compost. It also increases the biological activity of the soil and stimulates plant growth.[5]

Since Indian soils are low in carbon content, the response of fertilizer is not of the desired level. Long-term soil fertility experiments conducted under the aegies of ICAR has proved that addition of 5 tonnes of organic manure per hectare each year enhanced use and efficiency of fertilizer and water by 20 per cent.

10.9 Vermiculture/Vermicomposting

10.9.1 Introduction

Vermiculture means artificial rearing or cultivation of worms (Earthworms). Vermicomposting is the method of composting with earthworms and the end product is called "Vermicompost"; it is a mesophilic process carried out by worms within the pile of the moist organic material which are active at 10–32°C but best range being between 25–32°C. The worms feed upon the decomposted organic matter through the gut and excrete; the excreta is called "Vermicast" which is rich in humus, nitrogen, potassium, phosphorus, microbial activity, plant growth factors, vitamins, antibiotics and enzymes such as protease, amylase, lipase, cellulase, chitinase etc. The enzymes continue disintegration of the organic matter excreted as vermicasts. Vermicast contains excreta, earthworm cocoons and undigested food; the same is also fortified with pest-repellence attributes.[6] In addition, pathogens present in the material to be composted are destroyed during the process of vermicomposting.[7] The population of a number of beneficial microbes also increases in the vermicompost; a 4-fold increase in bacterial count and nitrogen fixing bacteria is reported. Actinomycetes, the microbes that create humus, increases 8 to 9 times compared to normal compost. The carbon to nitrogen ratio of the vermicompost is also less than the matter to be composted. This may be due to the combustion of carbon during respiration of the earthworms. Hence vermicompost is a storehouse of well-balanced fertilizer which provides plants with higher amounts of nitrogen, phosphorus, potassium and humus etc. The food for earthworms is partly decomposed materials like cattle dung, rumen contents, farm wastes, leaves, twigs, weeds, household garbage, organic municipal wastes, vegetable peelings and other organic materials.

10.9.2. Earthworms

Having learnt about the usefulness of earthworms, it is considered appropriate that a brief discussion is given about their life.

(1) There are many types of worms like roundworms, ribbonworms, flatworms and segmented worms. The earthworms is a type of segmented worm. There are about 2,700 species of earthworms around the world and live wherever soil is available. A common one amongst the segmented worms is "brandling" which is used for composting.

(2) Earthworms are fat and wiggly like our fingers (Figure 10.4).

(3) Earthworms has a mouth which is pointed (right-side in Figure 10.4) and the tail. They do not have teeth, eyes, ears

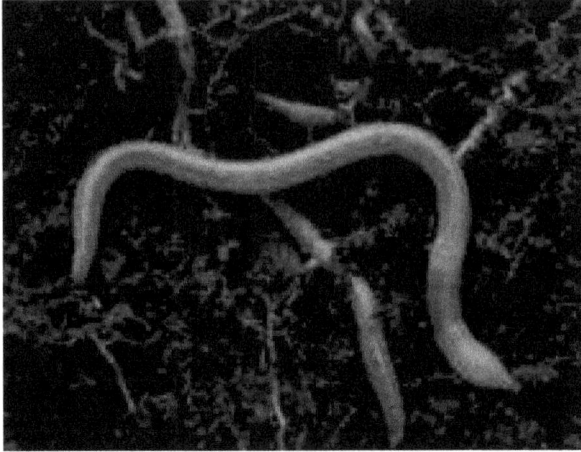

Fig. 10.4 : An Earthworm

or nose. They do have a brain, 5 hearts and parts inside their body which help them breath. Eventhough they do not have ears, they do not feel sound.

(4) The skin of earthworms need to be moist in order to breath. That is because they breath through their skin. If an earthworm dries out, it will die.

(5) Earthworm's body is covered with chemoreceptors which help the worm to taste things. Chemoreceptors are tiny sense organs which detect chemicals in the soil.

(6) Earthworms are invertebrates which means they do not have backbones.

(7) Earthworms are brown, pink or even red in colour. In the Philippines, however, there are blue earthworms and one kind in the United Kingdom is green.

(8) Earthworms are food for animals such as lizards, red ants, big black ants, centipedes, frogs, moles, turtles, snakes, birds etc. (Figure 10.5)

(9) The smallest earthworms are about 1 centimetre long and the largest one can be 4 metres long. One of the largest earthworms is called *Giant Gippsland Earthworm* found in one part of Australia.

Fig. 10.5 : Kingfisher with an earthworm as its prey

(10) If an earthworms breaks off close to the head or close to the tail, it will not die. Instead the bigger part will grow another head or tail as the case may be; the shorter part will die.

(11) Earthworms have been known to live more than ten years but most of them live less than one year.

(12) Earthworms are hermaphrodites which means that each earthworms has male parts that produce sperm and female parts that produce eggs. Thereby, all earthworms lay eggs.

(13) When two adult earthworms mate, they lie together and cover themselves in sticky mucus and pass sperm into each others body. The sperm makes the eggs inside each earthworm's body grow. Earthworms are able to mate when they are 12 months old.

(14) A thick ring of slime callsed "*Clitellum*" (Figure 10.6) forms around each adult earthworm's body. As the earthworm wriggle forward, the ring also moves forward, collecting eggs and slips over worm's head. As it slides off, the ends seal and form a cocoon.

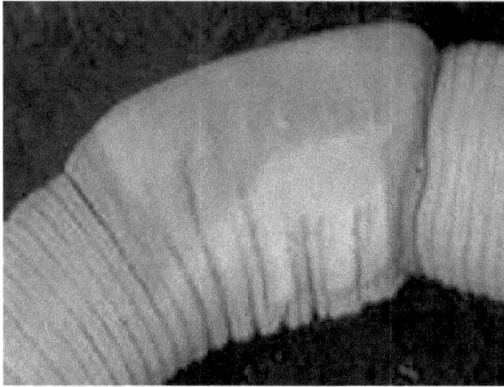

Fig. 10.6 : Clitellum around the worm's body

(15)   In warm weather, the eggs hatch after two weeks; in cold climate, however, they take upto three months to hatch. There are between one and twenty eggs in one cocoon but often only one or two eggs hatch. The baby earthworms are thin and white.

(16)   Earthworms are wonderful diggers. They dig passageways and burrows in the soil which let the water and air into the soil. This helps stop erosion and lets the water and air infiltrate to the roots of plants.

(17)   Earthworms live in cool, dark and damp places and eat dirt and rotting organic matter; inside the body, the food changes and comes out of their tail in the form of castings.

(18)   During winter, earthworms go below the frost-line at comfortable temperature and curl up in their burrows.

(19)   Earthworm have top and belly sides. Belly side has many bristles, called *setae* and is rough and paler in colour than top side.

For more information on earthworms, one is advised to go to the source.[8]

10.9.3 Species of Earthworms Commonly Used in Vermicomposting

The worms which are commonly used for vermicomposting invariably belong to the epigeic category. This is for the fact that in nature epigeic worms live in the top soil and duff layer on the soil surface. They can only move though the crevices of the surface and cannot make burrows in the soil. These are small, deeply pigmented worms and have poor burrowing capacity.

Epigeic worms are used in vermicomposting because their ideal environment can be duplicated in a bin or bed. They are voracious processors of organic material, do well in high-density culture and are extremely tolerant to the wide range of environmental conditions and fluctuations. Average weight of worms is about 0.5 to 0.6 gms but the adult ones weigh little over one gm each. Each worm eats roughly equivalent to its body weight in a day and produces roughly around 50 per cent of casts of the waste consumed; the volume gets reduced by 40–60 per cent of the organic waste consumed; the moisture of the caste ranging between 32–66 per cent and pH being around 7.0.

Eisenia foetida and *Eudrilus eugeniae* belong to the epigeic group of worms. The former is smaller in size as compared to the latter. These are purple red in colour and so are called red worms. *E. eugeniae* has been reported to have surpassed both in feeding as well as reproductive rates compared to other species of earthworms, for the reduction of urban and animal wastes. However, *E. eugeniae* cannot survive even brief periods at temperatures below 7.3°C (45°F). For this reason, *E. foetida* is used in preference over *E. eugeniae* where temperatures are likely to fall below 7.3°C.

Perionyx excavatus is an endemic worm and feeds actively on organic matter high in nitrogen. This apart, *Lampito mauritii* is also used for the purpose.

The species which are predominant in India, are *E. foetida, E. eugeniae* and *P. excavatus.*

Though not used for vermiculture, mention need to be made of Endogeic and Anecic categories of earthworms. Both are subsoil dewellers and Anecic may burrow as deep as 30 cm. Depending upon the species, they may burrow vertically or horizontally and even making very complicated network as in the case with Anecic species. They ingest organic and other material and help in the formation of organic-mineral complexes. The burrows loosen the soil which facilitate in the transport of water, air and spread of roots in the sub-soil.

10.9.4 Construction of Pit/Tank

For the preparation of vermicompost, one needs a storage place like a pit or a single tank or combination of two or four tanks where both earthworms as well as culture media can be stored together over a period of time under conditions suitable for preparation of the compost.

Pit and single tank methods are very easy and simple to be practiced and are the ones proposed to be discussed here.

As a general rule, vermicomposting should be carried out under shed and at an elevated place. Shed can be a thatched roof of suitable size and height. Elevated place is essential to avoid flooding of the facility during rains.

All the working details for a pit or tank for making vermicompost are same except that the pit is dug underground and tank is constructed over the ground.

Most convenient size of the tank from the point of working seems to be 1m (width) x 3m (length) x 0.60m (height). It can be constructed on unpaved ground using brick and cement mortar as shown in Figure 10.7. This will hold a mix of about 1,500 kg of dung and rumen contents and other organic wastes. To provide

Fig. 10.7 : Tanks for Vermicomposting

cross ventilation, it is desirable that the tank is provided with holes of 15mm diameter all around (walls) at 300 mm apart. The holes should be blocked with nylon screen (100 mesh) to prevent escape of worms form the pits.

Both the tanks as well as thatched roof are constructed in East-West direction lengthwise to protect the site from direct sunlight as well as rain.

10.9.5 Preparation of Vermicompost

Process details are as follows:

1. Spread organic matter like dry leaves, weeds, grasses etc. at the bottom of the tank about 5cm in thickness.

2. Take a mix of about 30–40 days rotten matter like dung, rumen contents etc. and some organic matter like dry leaves, vegetable peelings etc. and spread the same over the bed as prepared under (1).

3. For best results, the bed as prepared under (2), should be allowed to rest for about 3–4 weeks. This is for the fact that just as under composting, the bed passes through mesophilic and thermophilic stages before it again returns back to mesophilic stage. The bed is now ready for inocculation with worms.

4. Put $2^1/_2$ kg of earthworms (*E. foetida*) on the surface of the mix as under (3) and cover with gunny bags. *E. eugeniae* can also be used along with *E. foetida* in places where temperatures do not fall below 7.3°C (inside the pile).

5. The worms spread all over the surface, start eating the mix and excrete; the excreta is called vermicast. As the mix is eaten, the worms go on burrowing further down and continue the process of eating and casting as they burrow down. Water may have to be sprinkled at the top every now and then to avoid drying and maintain proper moisture content of the mix of about 50 to 60 per cent.

6. Remove the gunny bags after about 10 days; a good top layer of vermicast must have been formed by now. If so, do not sprinkle water for at least 2 days, but cover back the tank with gunny bags. The top layer of vermicast will

partially dry during 2 days after which it is removed. The process of removal of vermicast continues until one reaches the bottom. By this time, the worms get accumulated at the bottom layer and infact have multiplied many more times. The process of vermicomposting is complete in about 2–3 months period.

Vermicasts are small granules which comfortably pass through 2mm sieve; the casts should be partially dried and packed in polythene bags at around 15–20 per cent moisture content. The colour of the vermicompost is something like the colour of powder of roasted coffee beans.

10.9.6 Advantages of Using Vermicompost

1. Vermicompost is a natural eco-friendly fertilizer prepared from biodegradable organic wastes and is free from chemical inputs; as such it does not have any adverse effect on soil, plant or environment.

2. It improves soil aeration, texture and tilth, thereby reducing soil compactation.

3. It improves water retention capacity of the soil because of its high humus content.

4. It promotes better root growth and nutrient absorption.

5. It improves nutrient status of soil - both of macro and micro nutrients.

10.10 Vermicomposting in Ground Heaps

Instead of pits, vermicompost can be prepared in ground heaps. Dome shaped beds are prepared on the ground with material to be composted in the same fashion as for pit composting; the length, breadth and height also being same. The heap is then inoculated with worms and covered with gunny begs for maintaining favourable temperature and moisture. All other working details being same as for pits.

10.11 Phosphorous Enriched Vermicompost

Phosphorous enriched or P-enriched vermicompost can be prepared with the addition of Jhabua rock phosphate (30–32 per cent P_2O_5) @ 2.5 per cent at the stage of vermicomposting to the material to be composted. All other working details remains to be the same as under vermicomposting.

It may be said that there are variations in the methods of vermicomposting reported by different workers. Some workers for example, prefer to add worms in layers at different depths instead of the top as suggested in this chapter, while others may use in combination like *E. foetida* and *P. excavatus* and so on. It is perhaps because of the differences in the nature of raw materials from place to place as well as the climatic conditions and ones own experience.

10.12   Vermiwash

Vermiwash is essentially mucus and other body fluids of activated earthworms released in water. The method of preparation of vermiwash as well as activation of worms is rather simple as described by Singh, A.K. *et al.*[6] One kilogram of earthworms devoid of their casts are released into a trough containing 500 ml of lukewarm distilled water (37°C to 40°C) and agitated for 2 minutes; the agitation and warmth of the water help the worms to get activated and in turn, release mucus and body fluids into the water. After this wash, the earthworms are taken out and given a second wash in another 500 ml of distilled water at room temperature (around 30°C). Thereafter, the worms can be taken out and released back for vermicomposting. The ordinary water wash serves two objectives, washes out left over mucus and body fluids on the one hand and helps the worms to revive back from the shock. Both the washes are mixed and the mix is called "Vermiwash".

Sathe, T.V. has described another method of preparation of vermiwash.[9] The method is somewhat different than the one given here; ordinary potable water has been used in this method replacing distilled water. This product is a liquid fertilizer and after dilution with water, used as fouler spray. When mixed with 10 per cent of diluted urine of cow, finds use as pesticide on certain agricultural crops. Vermiwash thereby, finds use both as fertilizer as well as pesticide which makes it an important component of modern vermiculture.[9]

10.13 Price

The value of vermicompost for use as manure has already been discussed in detail; because of its innumerable virtues, it sells at around 3–5 times the price of compost. The retail price of vermicompost being around Rs. 10 per kg and that of compost Rs. 3 per kg in metro cities in India. Wherever possible, thereby, organic materials can better be converted into vermicompost.

10.14 Composting vs Global Warming

The heaps of dung, urine, garbage, sweepings and other organic matter for their conversion into compost may not be a very pleasing sight in most of the rural India but it may be surprising that this is the nature's way of prevention of global warming and thereby saving the mother earth from catastroph which is feared to wipe out much of the presently existing fauna and flora.

When organic matter of any description is burnt, most of the carbon as present (in the organic matter), gets converted into carbon dioxide (CO_2). Similarly, organic matter trapped under damp and marshy lands in anaerobic conditions, decomposes forming methane (CH_4), carbon dioxide and other gases in small quantities. All these gases slowly creep out into the atmosphere. Methane, no doubt, can be used as fuel but that is possible only when produced in economically viable quantities and controlled conditions.

Both carbon dioxide as well as methane (both are carbon compounds), as is well known, contribute to global warming. Infact, methane remains in the atmosphere for approximately 9–15 years and is over 20 times more effective in trapping heat in the atmosphere than the carbon dioxide over a period of 100 years. [10]

Composting eliminates the formation of carbon dioxide or methane (both gases do form but in very small quantities) and thereby arrests carbon from going back into the atmosphere. During the composting process, most of the carbon available in the organic matter is utilized by the microbes for their growth and multiplication as has earlier been discussed. And finally after their life cycle and death, the very microbes in the compost become the source of supply of carbon and other desirable nutrients as needed for growth and health of the plant.

Composting, thereby is not only desirable for agriculture, plant growth, food and feed supply, prevention of environmental pollution, control of many diseases but will also contribute significantly in the prevention of global warming. And that is how nature works silently, all the 24 hours, and making our lives pleasant and comfortable. Oh, wonderful nature.

References

1. Lynch, J.M. and Poole, N.J (1979) : *Microbial Ecology – a conceptual approach*. Blackwell Scientific Publications, Blackwell.

2. Lal, L. and Gupta, D.K (2008) : *Composting Technology*. Agrotech Publishing Academy, Udaipur, Rajasthan.

3. Mann, I (1962) : *Processing and utilization of animal byproducts* (FAO Agricultural Development Paper No. 75). Food and Agricultural Organisation of the United Nations, Rome.

4. Khanna, S.S (Jan. 2010) : Personal discussions. 323, Krishi Apartments, Vikas Puri, Block D, New Delhi.

5. Han, Y.B (1978) : Microbial utilization of straw – a review. *Advances in Applied Microbiology* 30, 145/46.

6. Singh, A.B., Manna, M.C., Ganguly, T.K. and Tripathi, A.K (2005): Vermicomposting. A technology for reclycling of organic wastes. Indian Institute of Soil Sciences, Bhopal (India.

7. Edwards, C.A. and Lofty, T.R (1977) : *Biology of Earthworms*. Chapman and Hall, London.

8. Earthworms (2010) : *www.kidcyber.com.au/topics/worms.htm.*

9. Sathe, T.V (2004) : *Vermiculture and Organic Farming*. Daya Publishing House, New Delhi - 110 035.

10. Adobe Reader_USEPA-Methane : http://www.epa.govt/egi-bin.epaprinting.egi.

CHAPTER 11

DISPOSAL OF CONDEMNED MATERIALS

11.1 Introduction

Condemned meat from a slaughterhouse will include live animals condemned on antemortem inspection and carcasses, parts of carcasses, organs, offals or tissues condemned on post-mortem inspection which have been declared unfit for human food by a meat inspector. Animals awaiting slaughter are condemned on antemortem inspection when they suffer from scheduled or infectious diseases which are communicable to man eg. anthrax, foot and mouth, rabies, tetanus, etc. Carcasses which are dangerous, or if they show deviation from the normal as to render them repugnant or not considered nutritious to the consumer are condemned on post mortem inspection. Parts of carcasses, organs, offals, etc are condemned on post mortem inspection if they show lesions due to disease or if they are injured eg. flesh of cattle and pigs affected with a very large number of cysticercus cysts (measly beaf and measly pork), organs and lymph nodes affected with acute tuberclosis, parts of liver and lungs affected with hydatid cysts, flesh affected with tumours, flesh of still born calves, condemned head, feet, liver, spleen, lungs, kidneys, heart, intestines, uterus, oesophagus etc.

Animals found dead on arrival at the slaughterhouse – death occurring due to disease, accident, suffocation and exhaustion are consigned as condemned meat (materials) and require suitable disposal.

11.2 Rules with Respect to Condemned Materials

Condemned animals and meats from the slaughterhouse and other sources, if handled with care is a valuable material and could be processed into meat meal and other useful products. The methods of disposal of animals suffering from scheduled diseases are governed by the animal diseases acts and rules of the governments of various states/countries. Wherever rules permit, it should be our aim to salvage as much of the condemned animals/meat as possible for processing into various products. From the point of view of disposal, the condemned animal/meat could be divided into two categories – low risk materials and high risk materials.

11.2.1 Low Risk Materials

Those materials which do not pose serious risks for man, livestock and environment in their handling could be considered as "low risk materials". Although the hazards presented by these materials may be moderate, their handling should be minimised as far as possible. Such materials are amenable to decontamination and conversion into useful products; for example, meat meal could be produced from these materials by the process of "rendering" which is discussed under chapter 3.

Carcasses of animals dead due to exhaustion and accident and meat condemned due to *C. bovis* and *C. Cellulosa* etc could also fall under the category of low risk materials.

11.2.2 High Risk Materials

Materials posing high risk to man, livestock and environment during their handling could be called as "high risk materials". Animals suffering from zoonotic diseases such as anthrax, brucellosis, foot and mouth, rinderpest, sheep pox etc may fall under this category. A zoonotic disease is one that primarily infects animals but can secondarily be transmitted to humans. Among these, the spores of anthrax and transmissible spongiform encephalopathy (TSE) are to be taken even more seriously. These pose great danger to man, livestock and environment with serious consequences. Both are briefly discussed.

Humans can acquire three types of anthrax, namely :

   i. Pulmonary form by inhalation of spores

  ii. Cutaneous form by contact of the spore with an open scratch on the skin, and

 iii. Gastro-intestinal form through consumption of anthrax infected meat. Anthrax spores are resistant to boiling and even when meat is boiled for long periods, the spores may remain viable. Consumption of such meat can expose the consumer to lethal gastro-intestinal anthrax.

Anthrax causes enlargement of the spleen and in advanced stages, there is a discharge of dark coloured blood from the mouth, nostrils and anus of the animal. In the case of pigs, anthrax lesions consist of localized gelatinous haemorrhagic exudate of the throat region and lymph nodes of the head.

Animals suffering from anthrax should be destroyed in their entirety at whatever the cost. On suspicion or identification, the animal should be isolated at great speed and slaughter of such animals is forbidden. If anthrax is recognised during slaughter and dressing, the carcass should be closed speedily ; the place where the slaughter and dressing took place, the tools used during slaughter and dressing, or handling and the materials which have come into contact with the infected carcass should be disinfected thoroughly and the carcass disposed off suitably. Disinfection could be done with boiling water, or 5 per cent caustic soda solution at 70° C for three hours.[1] Other methods of disinfection are also available.[2]

There are several forms of TSEs in different species of animals (Sheep, goat, deer, elk, mink) and humans. The one found in sheep and goat called "scrapie" known for more than 200 years but there are no records of transmision of scrapie to humans. In 1996, the identification of a new form of Creutzdt-Jacob Disease (CJD) in young people in U.K. raised the concern that causative agent for Bovine Spongiform Encephalopathy (BSE or also called mad cow diease) had transmitted from cattle to humans.

TSEs are a group of rare neurodegenerative diseases, sometimes called prion disease.

11.3 Destruction Methods

There are number of methods which are followed for the destruction of condemned materials, which being

1. Incineration
2. Open air burning
3. Fixed facility incineration
4. Air curtain incineration

5. Alkaline hydrolysis

6. Anaerobic digestion

7. Landfill

8. Burial

Apart from the methods listed under 11.3, the method of rendering discussed under the chapter on rendering, is safely followed for processing of low risk materials.[4] Each of the methods listed will be discussed in brief.

11.3.1 Incineration

Incineration is a method of burning or cremation carried at temperatures between 800° C to 1000° C. This is the safest method of destruction as organisms of all types are completely destroyed. However, it may be difficult to burn whole carcasses, especially of large animals, without an appropriate oven called "incinerator". It may not always be possible to install an incinerator for various reasons. If such burning is not possible either due to the nonavailability of an incinerator or lack of wood or oil, the carcass may have to be buried.

11.3.2 Open Air Burning

The method can be inexpensive but has potential for environmental contamination.

11.3.3 Fixed Facility Burning

This procedure is biosecure but is expensive and has limited capacity.

11.3.4 Air Circulation Incineration

It is a mobile facility but is fuel intensive and requires experienced personnel to operate.

11.3.5 Alkaline Hydrolysis

The proces is carried out in stainless steel pressure vessel with added sodium hydroxide (caustic soda) at elevated temperatures (150°C) and extended time; it is known to destroy all kinds of pathogens and prions, but the technology is currently expensive and limited in capacity. As it destroys even prions, the method is particularly suited to carcasses or parts thereof, affected with TSEs. The resulting two products viz., liquid hydrolyzate and inorganic part bone powder, find many uses.

11.3.6 Anaerobic Digestion

In this method, the carcasses or other offals are stored in closed pits or closed tanks, so that the materials are allowed to decompose in the absence of air (oxygen). Treatment of large carcasses is not practical by this method. In Israel, poultry offals are often disposed off in concrete pits with concrete covers. Use of concrete pits prevents pollution of underground water.

The method suffers from certain disadvantage; anaerobic digestion creates very disagreeable odours and takes very long time for complete disintegration of the material. The harmful organisms may also stay viable for long periods.

These measures are absolutely necessary for containing the disease as spores of "Bacillus anthracis" could survive for years.

11.3.7 Landfill

It is a convenient and affordable method of carcass disposal but environmental and regulatory considerations may make this option less feasible, especially if an infectious material is involved.

11.3.8 Burial

Two methods have been recommended for burial of anthrax infected carcasses; same methods have to be followed for the destruction of other condemned materials, if such materials have not been used otherwise.

According to one method, anthrax infected carcass has to be buried deep at a depth of not less than 1.83 m from the surface of the ground and covered all around with 0.3 m of freshly slaked lime.[3]

According to the second method, the carcass must be buried in hole not less than 3.3 m deep, covered with lime or some other disinfectant, and then hole filled with stones and soil.[1]

Irrespective of the method of burial followed, the following general measures are to be taken and ensured.

1. Burial place should be located on elevated, dry ground and not in a water logged area.

2. The place should be registered, indicated with a sign board and protected (fencing). This is necessary as carnivorous animals may dig out the carcass to feed upon the same.

3. People and animals should be kept away from the site.

4. Tillage of land and lifting of drinking water should not be allowed from the areas around the site.

A very detailed review of the subject has been recently published.[5] Ideally, most reliable methods for ruminal carcass disposal are :

(i) incineration and

(ii) high temperature extended time alkaline hydrolysis.

However, decision for the method to be opted will depend on the type of material, environmental considerations, bio-security, availability of resources, public concern and the local laws as applicable for disposal.

References

1. Mann, I (1984) : *Guidelines on small slaughterhouses and meat hygiene for developing countries (VPH/83.56). World Health Organisation, Geneva.*

2. Polyakov, A.A (1975) : *Veterinary disinfection.* Kolos, Moscow.

3. Martin, C.R.A (1959) : *Practical Food Inspection, 5th Edn.* H.K. Lewis & Co., London.

4. Pearl, G.P (2009) : Zoonotic diseases – their relationship to rendering. Render, April 2009 (Page 58).

5. Vanier, M. *et al.* (2009) : Ruminant carcass disposal options for routine and catastrophic mortality. CAST (Council for Agricultural Science and Technology) issue Paper 41. www.cast_science.org

CHAPTER 12

TREATMENT OF SLAUGHTERHOUSE EFFLUENTS

12.1 Introduction

Effluents could be defined as waste waters originating from a human activity or an industry. Domestic sewage, for example, could be called as domestic effluents. Similarly, waste waters coming out from a slaughterhouse and a rendering plant are called slaughterhouse effluents and rendering effluents respectively. The nature of these two effluents is very similar and hence in future discussions even if slaughterhouse effluents only are mentioned, it will be applicable to rendering effluents as well with adjustments as appropriate.

Chemically, slaughterhouse effluents are similar to domestic sewage but may be more concentrated. They are almost wholly organic, chiefly dissolved and suspended material.[1] As such, the methods applicable for the treatment of sewage waters are found to be suitable for the slaughterhouse effluents as well, with suitable adjustments as necessary. Considerable information, both theoretical as well as practical, is available on the treatment of sewage. As such, while dealing with the subject of slaughterhouse effluent treatment, some references have been made to the work connected with sewage treatment; this information could be considered as applicable to slaughterhouse effluents as well.

Before dealing with the suggested methods of effluents treatment, for developing countries in particular, it has been considered necessary to briefly explain various facets of the problem such as types of pollutants, nature of slaughterhouse effluents, pollution caused by these effluents, Biological Oxygen Demand (BOD), Chemical Oxygen Demand (COD), microbes and their role in effluent treatment and so on. This will help in better and fuller understanding of the subject.

12.2 Pollutants

Pollutants are substances in the effluents which are actually responsible for the pollution of the environment. Pollutants may be present in soluble, suspended or colloidal form in the effluents.

The nature of pollutants present varies greatly and is dependent upon the very nature of the processes in the industry. A chemical industry, for example, may discharge more chemicals in the effluents as compared to food processing industry which discharges mostly biologically degradable organic matter. Pollutants belong to the following three categories.

1. Biodegradable
2. Non-biodegradable (Refractory)
3. Biologically accumulative

12.2.1 Biodegradable

A number of complex, organic substances fall under the category of biodegradable pollutants; they are so called, as they undergo biological degradation through the action of various decomposers (bacteria, fungi, etc.), yielding simpler, inorganic products such as CO_2, nitrates, sulphates, etc. as the end products of their activity. Microbial population consisting of a number of organisms as coliforms, pathogens, etc. as normally associated with biodegradable pollutants, are also reduced over the period due to natural die off; after death, they also become biologically degradable matter. When dead, insects, fish and higher animals also behave as degradable organic matter.

12.2.2 Non-biodegradable

Many substances are inert to biological action and are termed as non-biodegradable or refractory substances. Inorganic substances like chlorides, salts, etc. fall under this category; some organics like BHC, DDT, ABS and various metals also fall under the group of non-biodegradables.

12.2.3 Biologically Accumulative

Some of the substances falling under the group of non-degradables persist in nature and accumulate in living organisms; such accumulations could prove dangerous and even fatal to those who consume them. Examples are, accumulation of arsenic, mercury, manganese, to mention a few, which accumulate in aquatic organisms and fish which are consumed.[1, 2, 3, 4, 5]

12.3. Sources of Wastes from Slaughterhouses

Slaughterhouse is principally a place for killing and dressing of animals for the production of hygienic meat for human consumption. While it is so, a number of associated functions are necessary for proper functioning of a slaughterhouse. A typical slaughterhouse may have the following departments, in developed countries in particular.

1. Lairage for resting of the stock
2. Slaughter floor or slaughterhouse
3. Tripe room
4. Hide/skin room
5. Chiller room
6. Meat cutting and processing room
7. Solid waste and blood disposal
8. Electrical light and power facility
9. Water supply
10. Effluents disposal
11. Meat market
12. Amenities and offices
13. Storm water drainage
14. Transport vehicles
15. Parks, trees, roads, etc.

As for the waste generation, other than killing floor wastes, there is a regular flow (of wastes) from some other departments as well; these are tabulated in Table 12.1. To keep the environment clean, all the available wastes, liquid or solid, will have to be suitably processed and disposed.

Table 12.1 : Sources of wastes in a slaughterhouse

Source	Waste
Lairage/Stock yard	Dung, urine
Killing floor	Blood
Hides/skins	Hair, dirt, soil
Stomach, intestine, rectum	Injesta, liquor, dung
Carcass dressing	Trimmings, fat tissues, blood
Byproducts – edible/inedible	Fat tissues, offals, trimmings, condemned meats
Rendering	Stick water or press liquor
Floor washings *	Hair, trimmings, blood, dirty waters

* Killing floor, meat market, chiller room, meat cutting room, rendering floor, etc.

While a typical slaughterhouse of today may have all the departments as listed above, some of them could actually be considred optional. In fact, in the context of the developing countries, neither it is necessary nor it is the practice to club facilities like chiller room, meat cutting and processing room, meat markets and rendering with a slaughterhouse. This is for the fact that there is a preference for the consumption of fresh meat in almost all the developing countries; practice of deboning the carcass and sale of meat at the slaughterhouse also does not exist. Moreover, most of the soft tissues are consumed as food in most of the developing countries. As a result, the profile of the effluents as well as the wastes as available in the developing countries, both quality wise and quantity wise, will be significantly different as compared to a typically organised abattoir in a developed country. This aspect has elaborately been discussed in Chapter 2.

It is, therefore, important that both the nature or quality of the effluents and the quantity originating from a particular slaughterhouse or rendering plant should be adequately established before an effluent plant is planned.

It will be outside the scope of this paper to deal with the details of construction of a slaughterhouse. However, considerable information is available on the subject in the form of guidelines, model designs, material of construction, location and site, facilities, etc. required for the construction of a slaughterhouse in developing countries. The reader is advised to refer to the literature on the subject.[6]

12.4 Nature of Slaughterhouse Effluents

Slaughterhouse as well as rendering plant effluents fall under the category of degradable effluents, as they are rich in biologically degradable matter like blood, urine, injesta, meat trimmings stick water and fat; they are practically devoid of non-degradable and biologically accumulative substances.

The total quantity of waste water let out and concentration of organic matter in it vary greatly in different establishments and depends to a large extent on the number and type of animals killed or type of materials rendered and the manner in which some of the products and byproducts are handled. As for example, in most of the developing countries, paunch or rumen contents are generally dry dumped (i.e. the semi-digested material from the rumen is emptied into a trolley and transported); but in a number of organised abattoirs in many developed countries, these are wet dumped, whereby the paunch contents are allowed to flow in a stream of water which will be passed over vibrating or rotating screens to separate the solubles and fines from the coarse material. While the coarse material may be separately transported, the solubles and fines become a part of the effluents. On the other hand, in most of the developing countries, blood is wasted and becomes a part of floor washings of the killing floor. This tremendously adds to the BOD of the effluents.

The composition of effluents from some slaughterhouses in India and the USA is given in Table 12.2.

Table 12.2 : Composition of Effluents from different establishments
(All figures except pH expressed in mg/l)

Parameter	Perambur (Madras cattle slaughter house)[7]	Delhi slaughter house[8]	USA[9]
pH	7.5	-	-
Total solids	26,560	9,025	4,100
Volatile solids	23,800	6,900	2,050
BOD	12,600	4,290	1,000
COD	22,600	6,320	-
Nitrogen	1,200	680	154
Phosphorous	450	300	-
Grease	375	2,540	-

As is clear from the table, the composition of effluents from one slaughterhouse to another, might vary greatly.

12.5 Pollution Caused by Slaughterhouse Effluents

The effects of pollution due to slaughterhouse or rendering plant effluents are many and varied; some of the most visible ones are listed below:

a. *Depletion of dissolved oxygen :* Oxygen is sparingly soluble in water; higher the temperature, lower the solubility. At the temperatures which prevail in the tropics, the river waters contain dissolved oxygen to an extent of about 8 to 9 mg/l, when saturated. If effluents are discharged into surface waters or streams, depletion of dissolved oxygen takes place because the organic pollutants in the effluents also require oxygen for their stabilization. Dissolved oxygen is most important for the growth of aquatic life – fish and microorganisms; the depletion of oxygen causes destruction of this life.

b. *Foul odours :* Proteins, fats and carbohydrates present in the effluents are amenable to fast biodegradation leading to foul odours.

c. *Public nuisance :* The effluents become the breeding ground for flies and mosquitoes and attract rodents causing public nuisance.

d. *Disease :* Some of the effluents carry a variety of pathogens such as salmonella, S.aureus, Bacillus, Clostridium, that may transmit diseases both to human beings and livestock.

e. *Loss of recreational and navigational waters :* The water bodies spoiled by the effluents become less attractive both for recreational and navigational purposes.

f. Eutrophication and eventual loss of water source, fall in land value, reduced agricultural output and increased water treatment costs.[2,3,4,5]

Very many pollution effects as discussed above, are a direct result of the imbalances caused to the elements of nature by the polluting agents. A brief discussion on the mechanism of pollution caused and definition of some of the "terms" used in connection with the treatment of effluents is desirable.

One of the most crucial effects of the biodegradable materials on environment is the demand for oxygen as required for their treatment and stabilization. Oxygen demand is expressed in two ways viz. Biological Oxygen Demand (BOD) and Chemical Oxygen Demand (COD).

12.5.1 Biological Oxygen Demand (BOD)

BOD is the quantity of dissolved oxygen in mg/l required for the oxidation of organic matter by microbial action. It is a measure of the strength of organic matter in terms of its ability to deplete dissolved oxygen in waste water. Thus, the BOD indicates the extent of oxygen consumed by the microbes for degradation of carbonaceous matter. As the pollutants in slaughterhouse and rendering plant effluents are mostly carbonaceous in nature, their (effluents) BOD is high. Generally, the standard test consists in measuring dissolved oxygen depletion at 20° C for 5 days.

12.5.2 Chemical Oxygen Demand (COD)

COD is the amount of oxygen, expressed in mg/l consumed under specific conditions in the chemical oxidation of organic and oxidisable inorganic matter contained in a waste water, corrected for the values of chlorides. Accordingly, organic matter is oxidised with an oxidant like potassium dichromate ($K_2Cr_2O_7$) in the presence of sulphuric acid and a catalyst (like Ag_2So_4) and the oxygen required for oxidation of the organic matter is calculated. Standard Methods for determining COD are available.[10]

As could be seen, there is a basic difference in BOD and COD in as much as the former indicates the oxygen demand resulting from bacterial activity while the later is the measure of the oxygen demand owing to chemical oxidation. All chemically oxidizable compounds are not necessarily biodegradable and, hence the two tests are not identical.

The COD values of many organic compounds are lower than total organic carbon (TOC) values but higher than BOD. The COD of slaughterhouse as well as rendering plant effluents, which contain predominantly readily biodegradable organics may not be very much higher than BOD. On the othe hand, COD and BOD might differ considerably for the effluents of some of the industries where the pollutants though predominantly organic in natue, but are not easily biodegradable. This happens in pulp and paper industry, where lignin – an organic pollutant, is not easily biodegradable.[2]

12.5.3 Nitrogen and Phosphorous

Like carbon, nitrogen and phosphorous are important constituents of all living matter. Through the process of "ammonification", the organic nitrogen gets converted to ammonia (NH_3). Ammonia is highly soluble in water and depending upon the pH of the water body, it is present in the soluble form as NH_3 or NH_4^+. Through the process of "nitrification", ammonia is converted into nitrate. The nitrification takes place in two steps; in the first step, ammonia gets oxidised to nitrite state which, in turn, gets converted to the final state of nitrate. This could be understood by the following two equations.

1. $2NH_4^+ + 3O_2 \rightarrow 2NO_2^- + 2H_2O + 4H^+$
2. $2NO_2^- + O_2 \rightarrow 2NO_3^-$

As such under certain conditions, nitrogen and phosphorous can lead to eutrophication in natural water bodies. Eutrophication essentially means increase in nutrient supply, such as nitrogen and phosphorous by natural or artificial means in a form that it increases the productivity of water and thereby brings about consequent changes in plant and animal life of the water body; such changes may affect the beauty and utility of the water body and in due course of time threaten its very existence.

This apart, nitrates can seep through the soil in the leachate and pollute ground waters. Excessive concentrations of nitrates in drinking water cause health problems like methymoglobaenemia.[2,3,4]

Phosphorous is present in living cells and other organic matter present in the effluents; the decomposition of organics (by decomposers) yields orthophosphates (soluble in water).

12.6 Microbes and Their Role in Effluent Treatment

All organic matter essentially contains carbonaceous (carbohydrate, cellulose, starch), nitrogenous (protein) and fatty material together with phosphorous, iron and various other trace elements. The role of microbes in the natural biological treatment of effluents essentially consists in the breakdown of the organic matter into simple and stable substances.

Essentially, there are three types of microbes *viz.* aerobes, anaerobes and facultative microbes which are responsible for the breakdown of the organic matter. Aerobic organisms thrive in the presence of dissolved oxygen, anaerobes or anaerobic microbes function in the absence of dissolved oxygen and utilize oxygen present in NO_3, SO_4, amino acids etc. Facultative microbes are active under either of the above conditions. Among each of these three types of microbes, there are also a number of species, each one being specific for a particular function. A brief description on the mechanism of breakdown of the organic matter will be helpful in better understanding of the subject.

12.6.1 Aerobic Decomposition

In aerobic decomposition, carbohydrates which could be expressed by the chemical formula CH_2O, are broken down by the aerobes to yield simple end products such as CO_2 and the energy as shown below; the energy in turn, is utilized by the aerobes for their own growth.

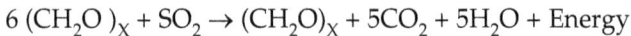

$$6\,(CH_2O)_X + SO_2 \rightarrow (CH_2O)_X + 5CO_2 + 5H_2O + \text{Energy}$$

The nitrogenous component present in the organic matter is generally first converted into ammonia by the aerobes, then to nitrites and finally to nitrates. Autotrophic bacteria such as nitrosomonas and nitrobacter and some heterotropic bacteria take part in the conversion of ammonia into nitrates as illustrated below.

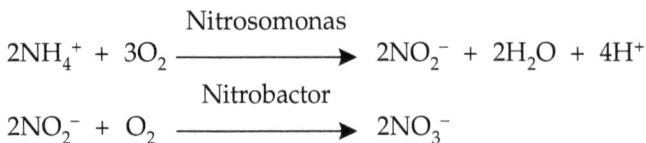

$$2NH_4^+ + 3O_2 \xrightarrow{\text{Nitrosomonas}} 2NO_2^- + 2H_2O + 4H^+$$

$$2NO_2^- + O_2 \xrightarrow{\text{Nitrobactor}} 2NO_3^-$$

In the aerobic decomposition, the sulphides present in the sulphurous material present in the organic matter, gets oxidised to stable sulphates. Here again, specialised microorganisms are responsible for the conversion of sulphides into sulphates.[2]

12.6.2 Anaerobic Decomposition

The anaerobic decomposition is comparatively more complex and not fully understood so far. A diagram showing the degradation of the organic matter to their component subunits and methane gas ultimately is shown in Fig. 12.1

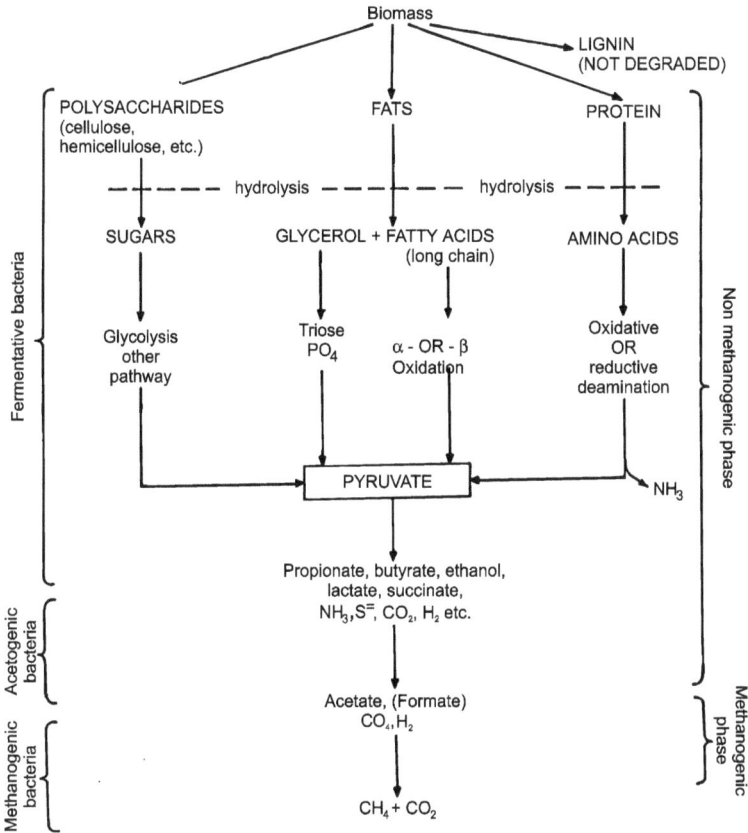

Fig. 12.1 : Flow diagram illustrating the major pathways for the degradation of major categories of organic compounds under anaerobic condition.

The carbonaceous matter is first converted into alcohols and volatile acids mainly acetic, propionic and butyric. This is demonstrated by the following equation in which only acetic acid is produced.

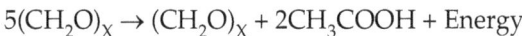

$$5(CH_2O)_X \rightarrow (CH_2O)_X + 2CH_3COOH + Energy$$

The acids are acted upon by the methanogens or methane bacteria to produce methane. A number of types of methanogens are present in the system – each being responsible for the conversion of the respective volatile acid into methane; conversion of acetic acid can be shown by the equation given below :

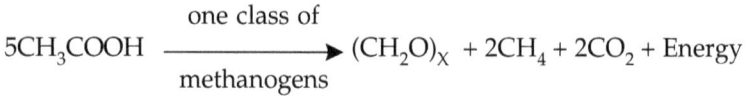

$$5CH_3COOH \xrightarrow[\text{methanogens}]{\text{one class of}} (CH_2O)_X + 2CH_4 + 2CO_2 + \text{Energy}$$

By another class of methane bacteria, propionic acid (CH_3CH_2COOH) is first converted into acetic acid, CO_2 and CH_4 and thereafter the conversion of acetic acid into CO_2 and CH_4 as shown above. Butyric acid (C_3H_7COOH) also yields CO_2 and CH_4 by another class of methanogens.

Nitrogenous material present in the organic matter, breaks down into ammonia or nitrogen itself and sulfurous material into sulphides (such as H_2S) in the anaerobic decomposition process.

The more resistant products of bacterial decomposition get converted into humic substances. Generally speaking, humic products are formed by the combination of aromatic compounds present in the organic matter, with the decomposition products of proteins and polysaccharides.

The various end products of the aerobic and anaerobic decomposition processes like CO_2, NH_3, sulphides, nitrites, etc. may undergo further changes under conditions of different ecosystems and finally get utilized as it is or in changed form, as source of nutrition for different kinds of living matter like plants, algae, zooplakton, worms, fish, animal and man himself. Algae for example, can grow on simple inorganic substances, like CO_2, NH_3, etc. in the presence of sunlight. Fuller discussion on these aspects will be outside the scope of this write up and one is advised to refer to the literature for further information.[2,3,4,5]

12.7 Aim of Effluent Treatment

The strength of effluents is expressed in terms of total solids while the volatile component of the total solids is known as volatile solids. The aim of any effluent treatment is basically its "stabilization" which essentially means reduction and/or biological degradation of volatile solids. During the course of effluent treatment, some sludge is formed; it is desirable that the sludge is obtained in a physical condition which will readily dewater under the influence of gravity. This property is of practical importance and helps in easy drying of the sludge on a sludge drying bed.

In addition, the aim is also intended to reduce or eliminate pathogens from the effluents during the course of their treatment.

12.8 Reduction of Effluents and Pollutants

It is important to point out that the volume of effluents generated and so also the load of pollutants in the effluents is greatly dependnt upon the practices of rendering plant and slaughter of the animal, dressing of the carcass and the manner in which the various byproducts generated are collected and utilized. Hence, every effort should be made to reduce the volume of the effluents as well as the load of pollutants in the same. This can be achieved by the following measures.

12.8.1 Dry Slaughtering

The use of water is reduced considerably in the so called practice of "dry slaughtering" when compared to modern methods of slaughter. Dry slaughtering is used to describe a process in which all operations of dressing the carcass viz. flaying, evisceration, splitting and despatching are done without the carcass coming into contact with water, either directly or through wet walls, floors or equipment.[11] It should not be understood that less use of water may lead to a less hygienic meat; on the contrary, the carcass dressed in such a manner has better keeping quality as there is no chance of the carcass getting contaminated through water used for washing the carcass in the modern practice.

Typical water requirements for dressing a carcass are given below.[6]

Cattle	-	1000 liters/animal
Small ruminants	-	100 liters/animal
Pig	-	450 liters/animal

In many country slaughterhouses in the UK, dry slaughtering is followed. Water requirements as prescribed by the Meat & Livestock Commission, for such slaughterhouses are as folows :

Cattle	-	260 liters/animal
Small ruminants	-	45 liters/animal
Pig	-	450 liters/animal

These amounts also include for wash down etc.[11]

One could appreciate that the amount of water used in the dry slaughtering method could be brought down to almost 30 per cent when compared to the modern practice. This helps in the reduction of effluents considerably.

In dry slaughtering, utmost care is needed in the dressing and handling of the carcass where, infact, many of the sophisticated equipments required for modern slaughtering can be done away with. Of course, time taken for dressing is comparatively little more than modern practices but this should not be a problem when small numbers of animals are slaughtered. Hence, the process is well suited for small and medium throughputs. With a number of advantages which the process offers such as low investment on slaughterhouse infrastructure, lower water consumption and production of hygienic meat, it merits attention for adoption in the developing countries.

12.8.2 Faeces, Blood, Injesta and Other Solids

Different kinds of wastes as available in a slaughterhouse are listed in Table 12.1. If allowed to be a part of the effluents, the load of the pollutants in the effluents will increase tremendously and hence the cost of treating the same. Every effort should therefore be made to reduce the load of pollutants to the minimum. Accordingly, the bulk pollutants should be collected separately and processed appropriately. This is possible. The injesta, for example, should be "dry dumpted" as explained under 12.4. Faeces can also be collected without any difficulty; the collection of blood and other solids (meat trimmings, etc.) is also possible. Methods of the treatment and processing of such bulk wastes have also been discussed in detail in different chapters.

12.9 Treatment and Disposal of Slaughterhouse Effluents

As already discussed, the slaughterhouse as well as rendering plant effluents are rich in varied kinds of organic matter, which may be in the form of solids, fats, suspended particles and solubles.

Before the final disposal of the effluents could be considered, the effluents have to be suitably treated. Essentially, the treatment aims at reducing the load of the pollutants in the effluents so that the disposal of the so treated effluents does not create environmental problems.

Different systems of treatment and disposal have been suggested based on the size of operations which have been broadly classified as follows :

1. Very small   - Slaughter of upto 2 large animals/day

2. Small   - Slaughter of upto 10 large animals/day

3. Medium   - Slaughter beyond 10 large animals/day

From the point of view of cost, simplicity, ease of operations and effectiveness of the system, the following four methods of treatment and disposal are suggested for adoption.

1. Very small slaughterhouse
                - Soakage pit

2. Small slaughterhouse
                - Septic tank followed with sub surface irrigation

3. Medium slaughterhouse
                1. Anaerobic lagoon followed by waste stabilization pond

                2. Anaerobic lagoon followed by aerated lagoon.

Irrespective of the process to be followed, it is always advisable to trap solids and grease from the effluents.

12.9.1 Screening of Solids

In small slaughterhouses, screening of solids can be easily effected by providing a vertical screen to the effluent discharging drain[12] from the slaughterhouse as shown in Fig. 12.2. The trapped solids are collected manually with the help of a rake.

12.9.2 Trapping of Grease

Slaughterhouse effluents always have some grease in them. This can come either in the form of small pieces as trimmings of fat tissues or melted by warm water used for washing hands, edible offals, tripe, etc. If not removed, the fats clog the soil and interfere with the effluent treatment systems. Small quantities of fat, however, do not pose very serious problem in the anaerobic method of treatment.

The fat can be trapped by installing grease trap to the drain as shown in Fig. 12.3. The grease trap is a simple holding tank where the fat has chance to congeal and float on the top, which is skimmed off and removed.

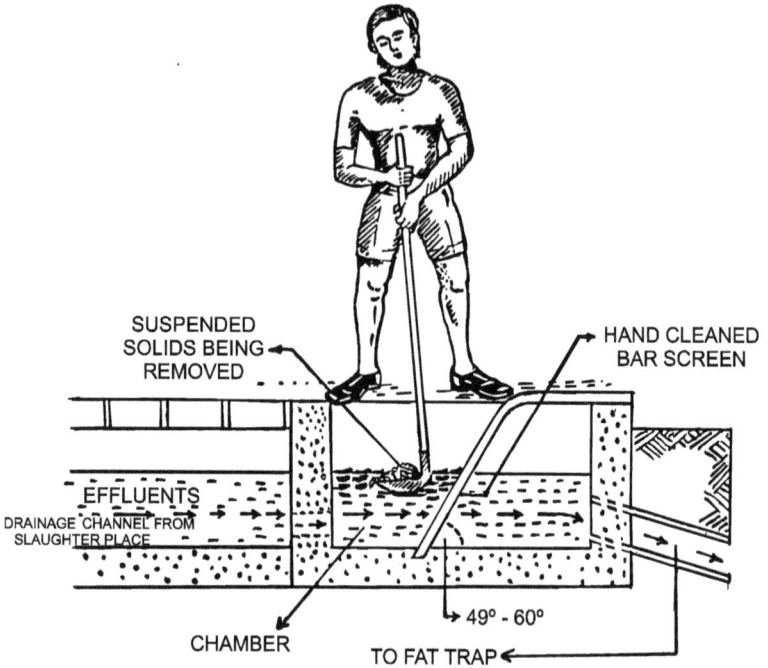

Fig. 12.2 : Screening and removal of solids from the effluents.

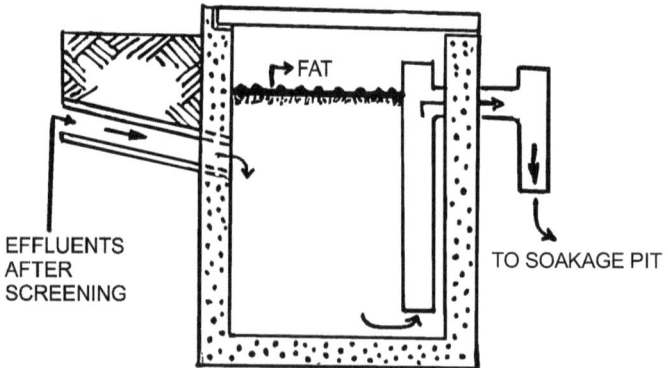

Fig. 12.3 : A fat trap.

12.10 Soakage Pit

Soakage pit, also known as seepage pit, is a covered pit designed to permit liquid wastes to seep into the surrounding soil.

As discussed earlier, this is a method which could be followed for the disposal of effluents for very small slaughterhouses, where the quantity of the effluents is limited and the pollutants load has been brought down by adopting measures discussed under 12.8.2 and prior treatment of the effluents described under 12.9.1 and 12.9.2.

This is a very simple method of disposal but will be successful only in places where soil conditions are suitable viz. free draining. A hard, impermeable soil, for instance, would be unsatisfactory. Hard soil does not allow the water to drain or permeate through. Soak pits will also not be suitable where rainfall and ground water table are high. A schematic flow sheet of the system is shown in Fig. 12.4.

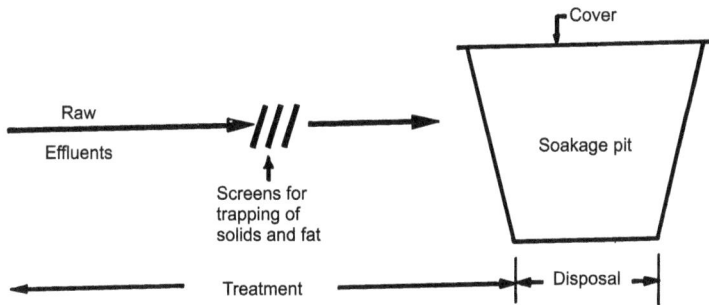

Fig. 12.4 : Suggested effluents treatment and disposal system for very small slaughterhouse

The suggested size for a soak pit could be 7 m deep and 2 m diameter. Over the time, soakage pits get blocked up due to continuous sludge formation and deposition of the same. Provision should, therfore, be made to dig additional pits as and when rate of soakage (seepage) declines.[12]

Since percolate from the pit can reach ground water aquifers, the risk of virus and pathogens transmission should be considered. Considerable work, both laboratory and at operational systems, has been done on viral movement from waste waters in Rapid Infiltration Systems; the results indicate minimum risk in general.[13]

12.11 Septic Tank Followed with Subsurface Irrigation

A flow sheet for the treatment and disposal by this system is shown in Fig. 12.5.

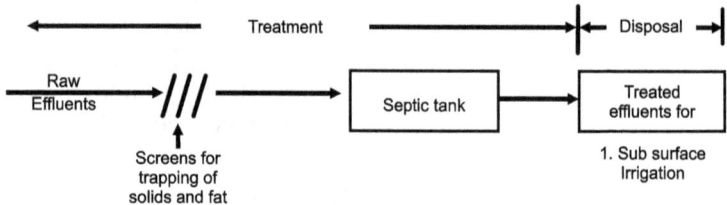

Fig. 12.5 : Suggested effluents treatment and disposal system for small slaughterhouse

A septic tank is a three compartment single storey settling tank in which the settled sludge is in immediate contact with the effluents flowing through the tank, while the organic solids are decomposed by anaerobic microbial action.

This is a good system where the treated effluents could either be disposed by using them for subsurface irrigation or in a soakage pit.

Should the treated effluents be used for subsurface irrigation, only non-root crop plants like banana, papaya, citrus, guava, mango, coconut, etc. or riverine species of trees should be grown; these waters should not be used for irrigating root crop plants like cassava, radish, carrot, potato, sweet potato and the like. There is a potential danger of communication of the pathogenic and viral diseases to humans and livestock through the use of root crops. For similar reasons, use of these waters for irrigating fodder and cereal crops should also be avoided.

Trenches approximately measuring 25 m length, 1 m width and 1.85 m depth, with a filling of upto 75 cm at the bed with large stones (100 cm & above) are recommended for irrigation.[6, 12] The trees are planted along the trenches as shown in Fig. 12.6. The trenches have been shown open in the Fig. with a view to get an idea about construction of the same. These should be suitably covered to avoid evaporation. Open trenches may also become breeding place for mosquitoes.

Fig. 12.6 : Subsurface irrigation trenches in parallel with trees planted along the bank.

The spacing between the trenches will depend upon the type of plants or trees chosen for planting; the recommended distance between trees like coconut and banana could be approximately 2.5 mtrs. These recommendations could, at best, be treated as guidelines and actual details could perhaps be decided upon according to local practice and in consultation with the local agriculture department.

12.12 Anaerobic Lagoon Followed by Waste Stabilization Pond

Anaerobic lagoons are useful where load of pollutants is high and are generally made use of for the treatment of effluents having BOD more than 1000 mg/l; heavy load (of pollutants) causes absence of aerobic zone in the pond. Anaerobic lagoons are usually 2.5 m to 5 m deep with a detention period varying between 10-20 days.

Complete treatment of effluents is possible using an anaerobic lagoon in conjunction with a waste stabilization pond. A case at Saint Helena, California where community waste waters amounting to about 1600 m³/day are treated using combination of such lagoons was reported.[14] Accordingly, the primary lagoon* *i.e.* anaerobic lagoon is 3 m deep, the detention time for the waste water being 20 days followed by 0.9 m deep, 10 day detention algal growth pond or waste stabilization pond (secondary pond**), which is allowed to be mixed at a velocity of 15 cm/sec. The results of treatment are shown in Table 12.3.

Table 12.3 : Removal of organic matter and nutrients in a combined system

Item	Percent removal in the combination system
BOD	97
COD	93
Carbon	78
Nitrogen	92
Phosphorous	64

It is possible to adopt the Saint Helena system for the treatment of slaughterhouse effluents with good results; the arrangement is shown in Fig. 12.7.

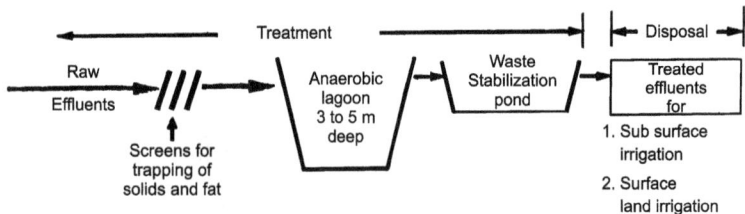

Fig. 12.7 : Suggested effluents treatment and disposal system for medium slaughterhouse (no mechanisation)

Depending upon the quality of the treated waters, these could be disposed off either for sub-surface irrigation or surface land irrigation.

---

* Meaning a pond or lagoon in which first treatment is given, irrespective of the category to which the lagoon belongs (aerobic, anaerobic, facultative).

** A pond which follows the primary pond.

12.13 Anaerobic Lagoon Followed by Aerated Lagoon

This system is precisely the same as that described under 12.12, except that the waste stabilization pond is replaced by the aerated lagoon as shown in Fig. 12.8.

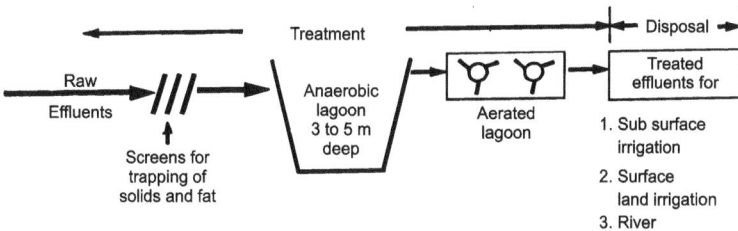

Fig. 12.8 : Suggested treatment and disposal system for medium slaughterhouse (partly mechanised)

An aerated lagoon is a basin in which effluents are treated on a flow-through basis; oxygen is supplied by means of surface aerators. The contents of the aerated lagoons are mixed completely and neither the incoming solids nor the biological solids produced from waste conversion settle out. The depth of the aerated lagoons varies between 2 to 3 m. Depending upon the climate, the detention time for the treatment of domestic sewage ranges between 3 to 12 days and can effect BOD removal from 70 to 90 per cent; energy costs are low ranging from 12 to 15 kwh/person/year.[2]

Aeration is carried out by the so called "vertical axis type" fixed or floating aerators. Mechanically aerated lagoons are simple and easy to operate.

Based on the experimental work done, this system viz. anaerobic lagoon followed by aerated lagoon had earlier been suggested for the treatment of slaughterhouse effluents.[7]

Depending upon the quality of the treated waters, these could be used either for sub-surface irrigation or surface land irrigation or even allowed to flow into the stream or river.

It needs to be emphasised that land requirement for natural effluent treatment systems is comparatively higher than to the many mechanised systems such as activated sludge process. However, they are very simple to construct – mostly amounting to earth work, easy to operate, no importation of costly equipment, almost negligible energy requirement and the personnel needed to run the system could be trained with ease and within a short-time. There could also be a reasonable income if the waters are used for irrigation as has been discussed under sub-surface irrigation. Natural systems are most effective under warm climates. This could perhaps be a coincidence that most of the developing countries are situated in the tropical and sub-tropical zones where climate is warm and rains seasonal. With this climatic advantage and with a number of other advantages offered by the natural systems as discussed, the developing countries should opt for these systems, even if the land requirements are comparatively higher. In fact, because of their environmentally friendly nature, these systems are gaining more and more popularity and acceptability even in the developed world.

Many factors such as pH, nature, composition and load of pollutants in the effluents, climate, condition of soil, terrain, degree of treatment desired and so on, have to be taken into consideration while designing a treatment system; discussion on the design aspects is outside the scope of this write-up. However, for complete information on these aspects one is advised to refer to the literature for details.[2,3,13,15] Consequently the recommended systems here should at best be taken as suggestive, in the sense that these are the most appropriate options for the developing countries. While designing a system, it may undergo necessary changes according to requirements.

References

1. Rudolf, W (1953) : *Industrial wastes.* Reinhold Publishing Corporation, New York.

2. Arceivala, S.J (1981) : *Wastewater treatment and disposal – engineering and ecology in pollution control.* Marcel Dekker, Inc. New York.

3. Woodard and Curran Inc. (Jan. 6, 2006) : *Industrial waste treatment handbook* (2nd edn.) Butterworth-Heinemann.

4. Andreoli, C.V., Sperling, M.V. and Fernandes, F (Jan. 5, 2007) : Sludge treatment and disposal. *Biological Wastewater Treatment,* Volume 6. IWA Publishing.

5. Baz, I.A. and Otterpohl, R (Jan. 8. 2008) : Efficient management of wastewater: Its treatment and reuse in water-scared countries. Springer.

6. *Standard design for small-scale modular slaughterhouses.* (1988) : (FAO Animal Production and Health paper No.73). Food & Agriculture Organisation of the United Nations, Rome.

7. Sastry, C.A., Kothandaraman, V and Murahari Rao, P (1972) : *Characterisation and treatment of slaughterhouse wastes.* Proc. Symp. Tannery and slaughterhouse waste Treatment, Central Leather Research Institute, Feb. 1-4, Madras.

8. Srivastava, S.K. and Seth, A.K (1972) : *Studies on the treatment of slaughterhouse waste using aerobic filter.* IBID, Feb. 1-4, Madras.

9. Southgate, B.A (1948) : *Treatment and disposal of industrial waste waters.* His Majesty's Stationery Office, London.

10. *Standard methods for examination of water supplies and waste waters.* APHA, 1971.

11. Goodman, J (1991) : *Private communication.* Meat and Livestock Commission, Milton Keynes, U.K.

12. Mann, I (1984) : *Guidelines on small slaughterhouses and meat hygiene for developing countries (VPH/83.56).* World Health Organisation, Geneva.

13. Reed, S.C., Middlebrooks, E.J. and Crites, R.W (1988) : *Natural systems for waste management and treatment.* McGraw-Hill Book Company, New York.

14. Oswald, W (1972) : *Complete waste treatment ponds. In:* Advances in Water Pollution Research, 6th International Conference, Jerusalem.

15. Eckenfilder, Jr., W.W., and Santhanam, C.J (1981) : *Sludge treatment.* Marcel Dekker, Inc. New York.

CONSOLIDATED STATEMENT OF PRODUCTS POSSIBLE TO MAKE FROM SLAUGHTERHOUSE WASTES

In the preceding Chapters, methods of production of various products from slaughterhouse wastes have been discussed; it is, however, possible to make many more products out of the very same raw materials. A consolidated statement providing details of the various raw materials, their approximate availability, products possible to be made and uses of such products, has been furnished in Table 13.1. The information could not be treated as exhaustive in the sense that many more products are possible to be made. It, however, amply demonstrates the variety and range of products which are possible to be made and their varied applications.

Table 13.1 : Products from Wastes of Animals Used for Food

Raw Material	Products	Uses

1. Hides/skins
  – 5% – 6% on live body
    weight of animal
	– Gelatine[1,2,3,4]
	
	– Leather
		– Edible/pharmaceutical
		
		– Leather products

2. Intestines (Small)
  – 50/80 ft/goat or sheep
  – 100/130 ft/cattle
  – 30/40 ft/pig
	– Sausage Casings[2,4,5,6,7]
	
	– Surgical catgut[5,7,8,9]
	
	– Sports guts[2,5,7]
	– Musical strings[2,5,7]
	– Other products[10]
		– Casing or container
		  for sausages.
		– Absorbable suturing
		  material used in surgery
		– Squash, tennis, badminton guts
		– Violin strings
		– Collagen sheet
		– Collagen tube, collagen powder
		  (Surgical Products)

3. Intestinal mucosa
  (obtained while
  cleaning intestines
  into sausage casings)
	– Heparin[2]
	
	– Fertilizer
	
	– Protein meal
		– Blood anti-coagulant
		  (Pharmaceutical product)
		– Organic fertilizer (Nitrogen source)
		– Livestock feeds
		  (Protein source)

4. Blood
  – 3.5% - 5.5% on live
  body Wt. of animal
	– Whole blood[11,12]
	– Blood albumin[11]
		– Blood sausages as food
		– Sausages
		– Leather finishing
		– Bakery Products

contd...

Table 13.1 – contd...

Raw Material	Products	Uses
5. Horns and Hoofs – 1% -2% on live body wt. of the animal	– Haemoglobin[11] – Other fractions[11] – Blood meal[2,5,6,13,14]	– Haemotonic – Pharmaceutical products – Livestock feeds (Protein source)
	– Raw horns/hoofs[2,5]	– Buttons – Combs – Cutlery handles – Fancy articles
	– Horn and Hoof meal[2,5]	– Organic fertilizer (Nitrogen source)
	– Hydrolysates[2,4,5,15,16]	– Flavouring agent (Soups) – Cosmetics (Hair care Products) – Livestock feed (Protein source) – Fire foam extinguisher
6. Bone – 15 % – 30% on live body wt. of the animal	– Raw bone (dry)[2,5] – Tallow[2,5,6,12,17,18]	– Handicrafts – Food – Livestock feeds – Industrial uses – Biodiesel
	– Crushed bone[2,5,6]	– Industrial uses (Ossein, glue, dicalcium phosphate, bone meal)

contd...

Table 13.1 – contd...

Raw Material	Products	Uses
	– Bone meal (raw)[2,5,6]	– Fertilizer (Phosphorous and Calcium)
	– Bone grist[2,5,6]	– Industrial uses (Bone char, bone glue manufacture)
	– Bone meal (steamed)[5,19,20,21]	– Mineral mixtures (Phosphrous and Calcium)
	– Bone char[2,5]	– Decolourising agent (Sugar/pharmaceutical industries)
	– Bone sinews[2,5]	– Glue manufacture
		– Fertilizer
		– Livestock feeds
	– Bone ash[2,5,6,22]	– Fertilizer
		– China porcelain
		– Mineral mixtures
		– Fertilizer
	– Dicalcium phosphate[1,19,20]	– Pharmaceuticals
		– Gelatine manufacture
	– Ossein[1]	(Edible/pharmaceutical/photographic).
7. Rumen Contents (RC) – 8% – 10% on live body wt. of the animal	– Composted RC[2,5,6,23]	– Manure
	– Dry RC[2,5,6,23,24]	– Manure
		– Livestock feed

contd...

Table 13.1 – contd...

Raw Material	Products	Uses
	– Ensiled RC[2,5,23,25,26]	– Livestock feed
	– Fresh RC[2,6,23,27,28]	– Bio-gas
		Manure
	– Vermicompost[23,29]	– Manure
	– Acid preserved RC[30]	– Livestock feed
8. Cattle tail hair* – 30-60 gm/cattle	– Dusters – Brushes[2,5,31,32]	– Dusters – Wall painting – Shoe polish – Carpet cleaning – Type writer cleaning etc.
9. Cow ear lobe hair* – Two pieces/cattle	– Artist brushes[2,5]	– Fine paintings
10. Bile liquid – 150-500 cc/cattle	– Cholic acid and its salts[2,33]	– Pharmaceuticals
	– Desoxy cholic acid and its salts[2,33]	– Pharmaceuticals
	– Cartisone[2,33]	– Pharmaceuticals

contd...

* other hairs like horse tail hair, goat hair, mangoose hair, deer hair, pony hair, squirrel hair etc. are all used for the manufacture of a variety of brushes.

Table 13.1 – contd...

Raw Material	Products	Uses
11. Gall stone	– Gall stone (dry)[2,33]	– Reported to be used in medicine and as lucky stone by the Orients.
12. Hide and skin trimmings	– Glue[2,4,5] – Gelatine[1,3,4] – Dog chews[2,5] – Collagen hydrolysates[34]	– Adhesive – Edible/Pharmaceutical – Dog treats – Cosmetics – Soft drinks – Weight reducing diets, soaps, etc.
13. Pancreas	– Pancreatin bate[2]	– Leather processing
14. Glands	– Glandular products[6]	– Pharmaceuticals
15. Soft tissues (Condemned meat, meat trimmings, uterus etc., fish wastes, poultry offals).	– Protein meals[2,5,23,35,36] – Fat[2,5,18,36]	– Livestock feeds (Protein source) – Industrial uses – Livestock feeds – Biodiesel
16. Poultry feathers	– Hydrolysed meal[37] – Hydrolysates[38,39]	– Livestock feeds (Protein source) – Cosmetics – Leather auxiliaries – Livestock feeds

contd...

Table 13.1 – contd...

Raw Material	Products	Uses
	– Raw feathers[5]	– Shuttle cock (badminton) – Pillows – Quilts
17. Fish bladder	– Raw bladder[34] – Soluble collagen[34] – Isinglass	– Soup – Cosmetics (Moisturising creams) – Clarification of wine/beer.

References

1. Gelatine (1984) : *An overvjew of the world market with special reference to the potential for developing countries.* International Trade Centre, YBCTAD/GATT, Geneva.

2. Mahendra Kumar (1989): *Handbook of rural technology for processing of animal byproducts.* (FAO Agricultural Services Bulletin No. 79). Food and Agriculture Organisation of the United Nations, Rome.

3. Ward, A.G. and Courts, A (1977) : *The science and technology of gelatine.* Academic Press, London.

4. *Proceedings and technical papers of the Symposium on utilisation of byproducts of leather industry (1960) :* Central Leather Research Institute, Madras.

5. Mahendra Kumar (1987) : *Processing of animal byproducts in developing countries - a manual.* Commonwealth Science Council, Marlborough House, Pall Mall, London SW1Y 5HX, UK.

6. Mann, I (1962): *Processing and utilization of animal byproducts.* (FAO Agricultural Development Paper No. 75). Food and Agriculture Organisation of the United Nations, Rome.

7. *Animal Casings (1973) :* International Trade Centre, UNCTAD/GATT, Geneva.

8. Barat, S.K. and Mahendra Kumar (1968) : *A process for the utilization of mammalian intestines for the manufacture of absorbable surgical catgut/suture/ligature "Plain and Chromic" or the like.* Ind. Pat. 116, 762, July 12.

9. Ranganayaki, M.D., Divakaran, S. and Barat, S.K (1972): *A study of physical and chemical changes during processing, tubing and sterilization of surgical sutures prepared by a patented process.* Leath, Sc., 19 (4), 103.

10. Mahendra Kumar (1980): *A process for the production of collagen sheet material from mammalian tissues.* Ind. pat. 154, 280, April 22.

11. Divakaran, S (1980): *Animal blood in food, feed, fertilizer, industry and laboratory.* NICLAI Publication – Central Leather Research Institute, Madras.

12. *Proceedings of training cum seminar on carcass and animal byproducts utilization (1981) :* Division of Livestock Products Technology, Indian Veterinary Research Institute, Izatnagar (U.P.), India.

13. Divakaran, S., Scaria, K.J. and Santappa, M (1978) : *Blood meal – its processing and utilization.* Leath. Sci., 25 (3), 127.

14. IS: 7060 - 1973: *Specification for blood meal as livestock feed.*

15. Hoshino, M., Tomita, S., Abe, K., Matsuda, Y., Terajima, K., Kojima, T., Sememiya, A., Hirose, T., Kira, K., and Namito, Y (1978): *Protein fire extinguishing liquors.* Jpn. Kokai Tokkyo Koho, *78, 135, 199, Nov. 25.*

16. Seiwa, Kesei K.K (1981): *Hair wave-setting preparations containing keratin hydrolysates.* Jpn. Kokai Tokkyo Koho 81. 118, 011 Sept. 16.

17. Divakaran, S., Ranganayaki., M.D.and Scaria. K. J (1971): *Utilization of slaughterhouse bones in India.* Khadi Gramodhyog, 21 (1), 91.

18. *Biodiesel* (May 31, 2009). http://en.wikpedia_org/Biodiesel.

19. IS. 1942- 1968: *Specification for Bone-meal, as livestock feed supplement.*

20. IS: 5672 - 1970 : *Specification for mineral mixtures for supplementing poultry feeds (reaffirmed 1980).*

21. IS: 1664 - 1981 : *Specification for mineral mixtures for supplementing cattle feeds (second revision).*

22. IS: 7061 -1973 : *Specification for calcined bone-meal as livestock feed supplement.*

23. Mahendra Kumar (2007): *Utilization of animal byproducts through semi-moist rendering.* Daya Publishing House, New Delhi.

24. Reddy, C.K. and Reddy, M.R (1977) : *Supplementation of dried rumen digesta in the rations of growing pigs.* Ind. J. Anim. Sci., 47 (4), 207.

25. Nicholson, J.W.G. and McQueen, R.E (1997): *Feeding value of ensiled mixtures containing rumen residue* (Abs.). J. Anim. Sci. 49 (Suppl.1) 138.

26. Rao, N. M. and Fontenot, J.P (1987): *Ensiling of rumen contents and blood with wheat straw.* Animal Feed Science and Technology. 18, 67.

27. Bepin Behari (1976): *Rural industrialistion in India.* Vikas Publishing House Pvt. Ltd., New Delhi.

28. Felix D. Maramba, Sr. : *Biogas and waste recycling – the Philippine experience.* Maya Farms Division, Liberty Flour Mills, Inc. Metro Manila, Philippines.

29. Edwards, C.A. and Lofty, T.R (1977) : *Biology of earthworms.* Chapman and Hall, London, U.K.

30. Gohl, B (1981): *Tropical Feeds.* (FAO Animal Production and Health Series No. 12, page 396). Food and Agriculture Organisation of the United Nations, Rome.

31. Merril Denison (1949): *Bristles and brushes.* Dodd, Mead and Company, New York.

32. *For quality performance, use a quality paint brush:* American Brush Manufacturers Association, 1990 Arch Street, Philadelphia, PA 19103.

33. *Byproducts of meat packing industry (Revised Edition) (1950) :* Prepared and edited by the committee of textbooks of American meat Inst., Institute of Meat Packing, University of Chicago, Chicago, ILL.

34. Mahendra Kumar (1984) : *Animal based raw material potential in India and their utilization,* Souvenir All India Bone Millers Association, No.202, Bhupindra Office Complex, 59, Rani Jhansi Road, New Delhi-110055.

35. Filstrup, P (1976): Handbook for the meat byproducts industry. Alfa-Lavel Slaughterhouse Byproducts Department, Titan Separator A/S, Danmark.

36. Bengtsson, O. and Holmqvist, O (1984) : *Byproducts from slaughtering – a short review.* Fleischwirtsch, 64 (3), 334.

37. *Feather meal :* National Renderers Association. Inc. 3150 Des Plaines Avenue, Des Plaines, Illinois 60018.

38. Sastry, T.P., Sehgal, P.K., Gupta, K.B. and Mahendra Kumar (1986): *Solubilized Keratins as a novel filler in the retanning of upper leathers.* Leath. Sci., 30 (12), 345.

39. Sehgal, P.K., Sastry, T.P. and Mahendra Kumar (1986): *Studies on solubilized keratins from poultry feathers.* Leath. Sci. 33 (12), 333.

INDEX

A

Abomasum, 88
ABS, 166
Aerobes (aerobic), 114
Anaerobes (anaerobic), 114
Ascomycetes, 137
Aspergillus, 138
Atlas Low Temperature Wet Rendering System, 23

B

B. stearothermophilus, 136
Bacillus, 113, 135, 170
Bacillus anthracis, 67, 113, 163
Bakshi ka Talab (BKT), 32
Basidiomycetes, 138
BHC, 166
Biogas from animal wastes, 21, 118-130
    animal wastes, 119
    biogas plant, 121-123
        digester, 123-124
        gas holder, 124
        inlet and outlet tanks, 124-125

composition of biogas, 120
composting, 129
general, 118-119
optimum parameters for anaerobic fermentation, 120-121
phases of biogas production, 121
products of biogas plant, 125
    biogas, 125-127
    liquid sludge, 128
    slurry, 127
    solid sludge, 127-128
toxicity of the slurry, 129
yield of biogas from different types of raw materials, 119-120
Biogas generation, 42
Blood, 73-85
    collection of blood, 73-74
    – – collection on floor, 78-80
    – – – – rail, 74-78
    introduction, 73

nutritional value of blood,
83-84

processing of blood, 80

– – absorbed blood, 81-83

– – blood meal, 80

– – lime treated blood, 80-81

sterilization, 83

Bone, 39, 58-72

availability of bones in
developing countries,
58-59

bone ash, 69-71

– meal, 63

bone meal as livestock
feed supplement,
67-68

– – -steamed, 68-69

– , meal, raw, 63-67

composition, 58

handicrafts, 60-61

dense and spongy
bones, 61-62

preparation of
handicrafts, 63

quality of finish, 63

treatment, 62-63

other bone based products,
72

uses, 71-72

– of bone, 59-60

yield, 58

Bone China, 71, 72

Bovine Spongiform
Encephalopathy, 161

C

C. bovis, 160

C. cellulose, 160

Central Leather Research
Institute, 31

Centri meal system, 23

Clitellum, 150

Clostridium tetani (tetanus
spores), 113

Clostridium, 67 , 113, 136

Collection, 33

Compost and horticulture, 39

Composting and
vermicomposting, 131-158

benefits of using compost as
manure, 147

composting-basic dyna-
mics, 133-135

composting-desirable
parameters, 131

atmospheric tempe-
rature, 132-133

carbon, nitrogen and
carbohydrates, 132

moisture, 132

pH, 132

porosity and aeration,
132

composting vs global
warming, 157-158

materials for composting,
139

methods of composting, 139

construction of bunkers,
140-141

first turning, 142

loading, 142

pit method of com-
posting, 139-140

second turning, 142-
143

stacking method of
composting, 140

microbes, 135

actinomycetes, 136-137

bacteria, 135-136

fungi, 137-138

protozoa, 138

phosphorous enriched
vermicompost, 155-156

price, 156-157

species of bacteria, 135

stack covered with vegetable
matter, 143

vermicomposting in ground
heaps, 155

vermiculture/vermi-
composting, 147

advantages of using
vermicompost, 155

construction of pit/
tank, 153-154

earthworms, 148-151

introduction, 147-148

preparation   of
vermicompost, 154-
155

species of earthworms
commonly used in
vermicomposting,
151-152

vermiwash, 156

windrow method, 143

culturing of inoculant,
144

introduction, 143-144

method of windrow
composting, 144-
147

Condemned materials, 159-164

Consolidated statement of
products possible to make
from   slaughterhouse
wastes, 187-196

D

DDT, 166

Department of Biotechnology, 32

Disposal   of   condemned
materials, 159-164

destruction methods, 161-
162

air circulation inci-
neration, 162

alkaline hydrolysis,
162-163

anaerobic digestion,
163

burial, 163-164

fixed facility burning,
162

incineration, 162

landfill, 163

open air burning, 162

introduction, 159

rules with respect to
condemned materials,
159-160

low risk materials, 160
high risk materials, 160-161
Dry meat meal, 36
Duke Continuous System, 23

E

E. coli, 92, 93
Eisenia foetida, 152, 154
Entrails, 35
Eudrilus eugeniae, 152

F

Flaying, 33
Food and Drug Administration (FDA), 114

G

General considerations, 1-11
　definition, 1
　earlier publications, 2
　future trends and suggested measures, 8
　　raising of more livestock, 9-10
　　strengthening of animal health care measures, 9
　　use of animal and other wastes, 8-9
　impact of non-utilization of animal offals, 4
　　communication of diseases, 4-5
　　economic loss, 5-7
　　pollution of environment, 4
　misleading concept, 3-4

perpetual wastage of animal byproducts, 7
　lack of awareness, 7
　– – technology, 7-8
　psychological bias, 7
scope of the technology, 1-2
source of byproducts, 2-3
Giant Gippslant Earthworm, 149
Government of India, 32

H

Humicola, 138

K

Keratin, 100
Kettle Rendering, 23
Khanna, S.S., 143

L

Lampito mauritii, 152

M

Maya Farms in Manila, 20
Maya Farms, 128
Mincing, 36, 50
Minor products, 15, 97-102
　gall bladder, 101-102
　glands and organs, 97-99
　hair, 100-101
　horns and hoofs, 100
　intestines, 99
　introduction, 97
Mucro, 137

O

Omasum, 88
Open Pan Rendering, 22

P

P. excavatus, 152
Pelleted feeds, 37
Pfaudler Conversion System, 23
Processing and Utilization of
    Animal Byproducts, 2
Protozoa, 135
Pseudomonas, 136

R

Religion, 2
Rendering, 22-57
    aims of rendering, 24-25
    choice of technology, 25-26
    definition, 22
    products of rendering, 24
    rendering process, 22-24
    semi-moist new technology
        package, 31-33
    – – rendering, 29
    – – – vs conventional
        rendering, 29-30
    – – unique features of the
        technology, 31
    simple cooking, 26-29
Reticulum, 88
Rumen contents, 86-96
    composition, 87-88
    definition, 86
    preparation of R.C. for
        livestock feed, 88
        acid preservation, 93
        ensiling, 90-93
        sun drying, 88-90
    profile of coarse and fine
        components of R.C., 93

R.C. as a useful raw
    material, 86-87
relation between volume
    and weight, 94-95
rumen-a living factory, 86
space for sun drying, 95
storage, 95
use, 94
yield, 88
Rumen, 88

S

Salmonella, 92, 93, 170
Sanitation and hygiene, 103-117
    building, 103-104
    dry slaughtering, 105-106
    education programme, 106-
        107
    general, 103
    means of communication,
        105
    microbes, 107-108
        cell structure, 109-110
        environmental factors,
            113
        moisture, 115
        molds, 116-117
        motility, 110
        multiplication and
            growth of bacteria,
            110-113
        nutrition, 113-114
        oxygen, 114
        pH, 114-115
        shape, 108-109
        size, 109

spore formation, 113
temperature, 115-116
viruses, 117
yeasts, 116
plant maintenance, 106
scheduled diseases, 107
transport vehicles, 104-105
water, 104
Semi-moist meat meal and its water activity, 52–54
Semi-moist meat meal, 24, 37
Semi-moist rendering vessel, 43
Semi-moist rendering, 8
Simple cooking, 23
Slaughterhouse effluents, 165-186
Solid urban wastes, 143
Stick water, 37
Streptomyces, 136

T

Thermoactinomyces, 137
Thermomonospora, 137
Thielavia, 138
Treatment and disposal of effluents, 40
Treatment of slaughterhouse effluents, 165-186
aim of effluent treatment, 175-176
anaerobic lagoon followed by
aerated lagoon, 184-185
waste stabilization pond, 182-183
introduction, 165-166

microbes and their role in effluent treatment, 172-173
aerobic decomposition, 173
anaerobic decomposition, 173-175
nature of slaughterhouse effluents, 169-170
pollutants, 166
biodegradable, 166
non-biodegradable, 166
– biologically accumulative, 167
pollution caused by slaughterhouse effluents, 170-171
biological oxygen demand, 171
chemical oxygen demand, 171-172
nitrogen and phosphorous, 172
reduction of effluents and pollutants, 176
dry slaughtering, 176-177
faeces, blood, injesta and other solids, 177
septic tank followed with subsurface irrigation, 181-182
soakage pit, 180
sources of wastes from slaughterhouses, 167-168

treatment and disposal of slaughterhouse effluents, 177-178

screening of solids, 178

trapping of grease, 178-179

Types of waste emanating from slaughterhouses: availability, 12

collection and disposal, 12-13

blood, 13

rumen contents, 13

bone, 13-14

horns and hoofs, 14-15

alimentary tract, 15

gall bladder, 16

soft and fat tissues, 16

keratinous fibres, 16-19

poultry offals, 19-20

effluents, 20-21

current practices of their collection, 12-21

disposal in developing countries, 12-21

U

United States of America, 114

USA, 22, 169

Use of semi-moist meat in pig feed, 37

Use of semi-moist meat meal in pig feed, 55

V

Vermicomposting, 131-158

W

Wet Rendering, 23

Windrow Method of Composting, 145

Z

Zygomycetes, 137